理工类地方本科院校新形态系列教材

Android移动开发技术

主　编　杨剑勇　钱振江

南京大学出版社

图书在版编目(CIP)数据

Android 移动开发技术 / 杨剑勇，钱振江主编. —
南京 ：南京大学出版社，2021.9
ISBN 978 - 7 - 305 - 24506 - 0

Ⅰ. ①A… Ⅱ. ①杨… ②钱… Ⅲ. ①移动终端—应用
程序—程序设计 Ⅳ. ①TN929.53

中国版本图书馆 CIP 数据核字(2021)第 103699 号

出版发行 南京大学出版社
社　　址 南京市汉口路 22 号　　邮　　编 210093
出 版 人 金鑫荣

书　　名 Android 移动开发技术
主　　编 杨剑勇　钱振江
责任编辑 吕家慧　　编辑热线 025 - 83597482

照　　排 南京开卷文化传媒有限公司
印　　刷 南京人民印刷厂有限责任公司
开　　本 787mm×1092mm 1/16 开 印张 12.5 字数 305 千
版　　次 2021 年 9 月第 1 版 2021 年 9 月第 1 次印刷
ISBN 978 - 7 - 305 - 24506 - 0
定　　价 39.80 元

网　　址：http://www.njupco.com
官方微博：http://weibo.com/njupco
微信服务号：NJUyuexue
销售咨询热线：(025)83594756

扫码可获取
本书相关资源

前 言

Android是Google公司开发的基于Linux的开源操作系统，主要应用于智能手机、平板电脑等移动设备。经过短短几年的发展，Android系统在全球得到了大规模推广，除智能手机和平板电脑外，还可用于穿戴设备、智能家居等领域。据不完全统计，Android系统已经占据了全球智能手机操作系统80%以上的份额，中国市场占有率更是高达90%以上。由于Android的迅速发展，市场对Android开发人才需求猛增，越来越多的人学习Android技术，以适应市场需求寻求更广阔的发展空间。

为什么要学习本书

市面上真正适合初学者的Android书籍并不多，为此，我们推出了本书供初学者使用。本书全部案例使用Android Studio开发工具，站在初学者的角度，知识讲解由浅入深，并采用当前最流行的案例驱动式教学，通过50多个案例来讲解Android基础知识，最后通过2个综合案例将全书的知识点进行串联综合。通过本书的学习，可以使初学者更好的掌握Android开发技术，为将来进一步地提高打下良好基础。

如何使用本书

Android开发使用Java语言。初学者在使用本书时，一定要具备Java基础知识，建议从头开始，循序渐进地学习，并且反复练习书中的案例，以达到熟能生巧的程度。如果是有基础的编程人员，可以选择感兴趣的章节跳跃式地学习，书中的案例最好自己动手实现代码。

本书共分13章，每章的具体内容：第1章讲解Android的基础知识，包括Android起源、Android体系结构等。第2章讲解Android开发环境的搭建、模拟器的配置、Android的项目结构和开发工具Android Studio的使用。第3章讲解界面设计、UI布局和项目资源等知识。第4章讲解Android开发中具备的一些控件知识。第5章讲解Adapter和AdapterView相关知识，将各种复杂的数据加载到控件上的方法。第6章讲解Activity，包括Activity创建、生命周期、数据传递等。第7章讲解Android中的重要的广播机制和内容提供者以及动态权限机制，使用广播接收者和服务实现后台程序。第8章讲解Android中的数据存储，包括文件存储、SharedPreferences等知识。第9章讲解多线程和服务。第10

章讲解 Android 中的网络编程，包括 HTTP 协议、HttpURLConnection、数据提交方式以及 JSON 和 GSON 的数据处理技术。第 11 章讲解使用百度地图 SDK 开发基于定位服务。第 12 章讲解基于本地的文件管理系统 App 程序。第 13 章讲解基于网络数据接口的综合 App 项目。

初学者在实战演练的过程中遇到困难，请不要纠结，继续往后学习，很多情况下，通过后面的知识讲解，前面不懂的技术就能理解了。除了多实践之外，多思考和多整理思路，认真分析和总结，也能帮助读者更有效率地掌握知识。

由于篇幅关系，程序尽可能展示框架结构，有些代码合并在一行之内显示，有些代码将用"……"代替，具体的代码实现请通过扫描随书的二维码查看。

致谢

本书的编写和整理工作由杨剑勇和钱振江完成，杨剑勇编写了第 1,4,6,8,12,13 章，钱振江编写了第 2,3,5,7,9,10,11 章。本书全部案例的编写和调试由两人共同完成。在本书的编撰过程中得到了常熟理工学院教务处和计算机学院领导的关心和大力支持。

意见反馈

尽管我们做了最大的努力，但书中难免会有不妥之处，欢迎各界专家和读者朋友们来信给予宝贵意见，我们将不胜感激。当您在阅读本书时，发现任何问题可以通过电子邮件与我们取得联系。

电子邮件地址：18013684925@qq.com

编者

2020 年 9 月

目　录

第 1 章

Android 简介

随着移动设备的不断普及发展，相关软件的开发也越来越受到程序员的青睐。目前移动开发领域中，以 Android 的发展最为迅猛，它推出短短几年时间，就撼动了诺基亚的霸主地位。通过其在线市场，程序员不仅能向全世界贡献自己的程序，还可以通过销售获得不菲的收入。本章将对手机操作系统、Android 发展史、Android 特性和 Android 平台架构进行介绍，让大家对 Android 有一个基本的了解。

1.1 移动平台系统

在手机问世的初期，很长一段时间内，都是没有智能操作系统的，所有的软件都是由手机生产商在设计时定制的。但是随着通信网络的不断改善，由早期的模拟通信网络（1G 网络），发展到广为使用的数字通信网络（2G 网络），到能方便访问互联网的第三代通信网络（3G 网络），再到现在正在使用的第四代通信网络（4G 网络），以及马上就要进入 5G 时代，手机已经不再只满足基本的通话、短信功能，而是逐步变为一个移动的 PC 终端，从而它也拥有了独立的操作系统。手机上的操作系统先后出现过 Windows Mobile、Windows Phone、BlackBerry OS、Symbian OS、Android 和 iOS 等，下面进行简单的介绍。

1.1.1 Windows Mobile

Windows Mobile 是微软公司推出的移动设备操作系统。由于其界面类似于计算机中使用的 Windows 操作系统，所以用户操作起来比较容易上手。它捆绑了一系列针对移动设备开发的应用软件，还预安装了 Office 和 IE 等常用软件，有很强的媒体播放能力。但是由于其对硬件要求较高，并且系统会经常出现死机的现象，限制了该操作系统的发展。

1.1.2 Windows Phone

Windows Phone 是微软公司 2010 年推出的新一代的移动操作系统。该系统与 Windows Mobile 有很大不同，它具有独特的“方格子”用户界面，并且增加了多点触控和重力感应功能，同时还集成了 Xbox Live 游戏和 Zune 音乐功能。

1.1.3 BlackBerry OS

BlackBerry OS（黑莓操作系统）是由加拿大的 RIM 公司推出的与黑莓手机配套使用的系统，它提供了手提电脑、文字短信、互联网传真、网页浏览，以及其他无线信息服务功能。其中，最主要的特色就是它支持电子邮件推送，邮件服务器主动将收到的邮件推送到用户的手持设备上，用户不必频繁地连接网络查看是否有新邮件。黑莓操作系统主要针对商务应用，因此具有很高的安全性和可靠性。

1.1.4 Symbian OS

Symbian OS(塞班操作系统)是一个实时性、多任务的纯32位操作系统。该操作系统功耗低、内存占用少,具有灵活的应用界面框架,并提供公开的API文档等,不但可以使开发人员快速地掌握关键技术,还可以让手机制造商推出不同界面的产品。但由于它只对手机制造商和其他合作伙伴开放核心代码,大大制约了它的发展。后来随着Android和iOS的迅速崛起,Symbian OS最终被替代。

1.1.5 iOS

iOS操作系统是苹果公司开发的移动操作系统,主要应用在iPhone、iPad、iPod touch以及Apple TV等产品上。它的屏幕是用户体验的核心,用户不仅可以在上面浏览优美的文字、图片和视频,也可通过多点触摸屏进行交互。另外,iOS允许系统界面根据屏幕的方向而改变方向,用户体验效果非常好。iOS使用Objective-C作为程序开发语言,苹果公司还提供了SDK,为iOS应用程序开发、测试、运行和调试提供工具。

1.1.6 Android

Android是Google(谷歌)公司发布的基于Linux内核的专门为移动设备开发的平台,其中包含了操作系统、中间件和核心应用等。Android是一个完全免费的手机平台,使用它不需要授权费,可以完全定制。由于Android的底层使用开源的Linux操作系统,同时开放了应用程序开发工具,这使所有程序开发人员都在统一的、开放的平台上进行开发,从而保证了Android应用程序的可移植性。Android使用Java作为程序开发语言,所以不少Java开发人员加入此开发阵营,这无疑加快了Android的发展。

1.2 Android移动开发系统

Android本义是指“机器人”,标志也是一个机器人。它是Google公司推出的一款开源免费的智能操作系统,其中包含了操作系统、中间件和核心应用等。

Android平台首要优势就是其开放性,允许任何移动终端厂商加入Android开发联盟,使其拥有更多的开发者。随着用户和应用的日益丰富,一个崭新的平台也将很快走向成熟。同时由于Android的开放性,众多的厂商会推出千奇百怪、各具功能特色的多种产品。功能上的差异和特色,并不会影响数据同步,再者Android平台提供给第三方开发商一个十分宽泛、自由的环境,而互联网巨头Google已经成为连接用户和互联网的重要纽带,Android平台手机可以结合这些优秀的Google服务。正是这些特性,使得Android系统受到如此欢迎,市场占有率超过百分之八十。

1.2.1 Android的发展历史

2003年10月,Andy Rubin等人一起创办了Android公司。2005年8月Google收购了这家公司,经过了数年的研发之后,2008年推出了Android系统的第一个版本。Google采用开放政策,任何手机厂商和个人都能免费获得Android系统的源码,且可以自由地使用,使得很多手机公司都推出了各自系列的 Android 手机，应用程序市场百花齐放。短短几年，就超过了已经霸占市场逾十年的诺基亚 Symbian，成为全球第一大智能手机操作系统。

Android 在发展的过程中，已经经历了十多个主要版本的变化，每个版本的代号都是以甜点名来命名的，该命名方法开始于 Android 1.5 版本，并按照首字母顺序：纸杯蛋糕、甜甜圈、松饼、冻酸奶、姜饼、蜂巢等。Android 迄今为止发布的主要版本如表 1－1 所示。

表 1－1 Android 主流版本、发布时间和对应的 API level

Android 版本名称	Android 版本	发布时间	对应 API
(no code name)	1.0	2008/9/23	API level 1
(no code name)	1.1	2009/2/2	API level 2
Cupcake	1.5	2009/4/17	API level 3，NDK 1
Donut	1.6	2009/9/15	API level 4，NDK 2
Eclair	2.0.1	2009/12/3	API level 6
Eclair	2.1	2010/1/12	API level 7，NDK3
Froyo	2.2.x	2010/1/12	API level 8，NDK 4
Gingerbread	2.3—2.3.2	2011/1/1	API level 9，NDK5
Gingerbread	2.3.3—2.3.7	2011/9/2	API level 10
Honeycomb	3.0	2011/2/24	API level 11
Honeycomb	3.1	2011/5/10	API level 12，NDK 6
Honeycomb	3.2.x	2011/7/15	API level 13
Ice Cream Sandwich	4.0.1—4.0.2	2011/10/19	API level 14，NDK 7
Ice Cream Sandwich	4.0.3—4.0.4	2012/2/6	API level 15，NDK 8
Jelly Bean	4.1	2012/6/28	API level 16
Jelly Bean	4.1.1	2012/6/28	API level 16
Jelly Bean	4.2—4.2.2	2012/11	API level 17
Jelly Bean	4.3	2013/7	API level 18
KitKat	4.4	2013/7/24	API level 19
Kitkat Watch	4.4W	2014/6	API level 20
Lollipop(Android L)	5.0/5.1	2014/6/25	API level 21/ 22
Marshmallow(Android M)	6.0	2015/5/28	API level 23
Nougat(Android N)	7.0	2016/5/18	API level 24
Nougat(Android N)	7.1	2016/12	API level 25
Oreo(Android O)	8.0	2017/8/22	API level 26
Oreo(Android O)	8.1	2017/12/5	API level 27
Pie (Android P)	9.0	2018/8/7	API level 28

1.2.2 Android 的体系架构

Android 系统采用了分层架构，共分为四层，分别是 Android 应用层、Android 应用框架层、Android 系统运行层和 Linux 内核层，如图 1-1 所示。下面分别介绍各层。

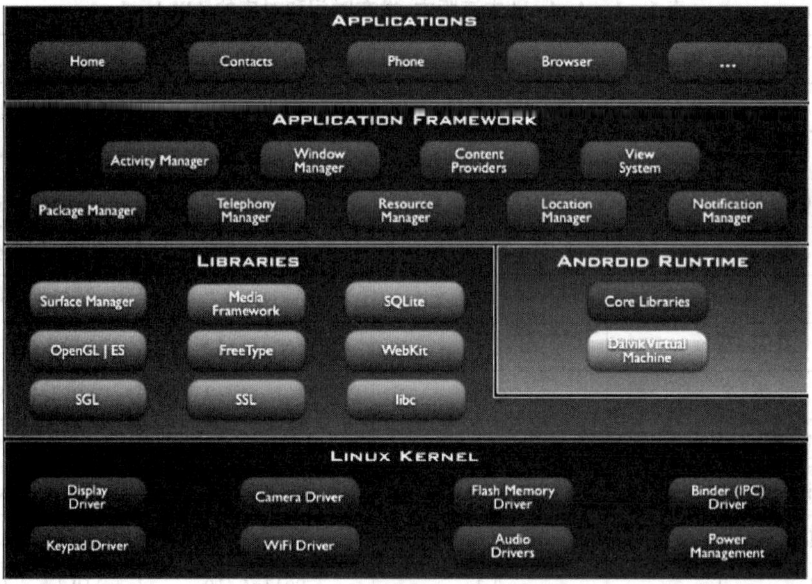

图 1-1 Android 的系统架构

1. *Linux 内核层*

Linux 内核层为各种硬件提供了驱动程序，如显示驱动、相机驱动、蓝牙驱动、电池管理等等。正是通过这些驱动程序来驱动设备上的硬件设备。

2. *Android 系统运行层*

这一层包括 Libraries 和 Android Runtime。

Libraries：核心类库包含了系统库和 Android 运行环境。系统库主要包括一组 C/C++ 库，用于 Android 系统中不同的组件，这些功能通过 Android 应用程序框架对开发者开放。下面是一些相关的核心类库：

(1) C 语言系统：派生于标准 C 语言系统，并根据嵌入式 Linux 设备进行调优。

(2) 多媒体库：基于 OpenCore 多媒体开源框架，支持多种视频、音频文件。

(3) 外观管理器：管理访问子系统的显示，将 2D 绘图与 3D 绘图进行显示上的合成。

(4) SGL：底层的 2D 图形引擎。

(5) OpenGL|ES：该库使用了硬件 3D 加速或高度优化的 3D 软件光栅。

(6) FreeType：用于位图和矢量字体的渲染。

(7) SQLite：一个强大的关系型数据库。

Android Runtime(ART)：Java 程序的运行需要 Java 核心包的支持，通过 JVM 虚拟机运行 Java 应用程序，Core Libraries 就相当于 Java 的 JDK，是运行 Android 应用程序所需要

的核心库，Dalvik Virtual Machine 就相当于 JVM，这是 Google 专为 Android 开发的运行 Android 应用程序所需的虚拟机。

Android 运行时使用 ART 虚拟机，区别之前的 Dalvik 虚拟机，ART 模式在启用后，系统在安装应用程序的时候会进行一次预编译，会先将代码转换为机器语言存储在本地，这样在运行程序时就不会每次都进行一次编译了，执行效率也大大提升。每个 Java 程序都运行在 ART 虚拟机上，该虚拟机专门针对移动设备进行了定制，每个应用都有自己的 Android Runtime 实例。

3. Android 应用框架层

这一层主要提供了构建应用程序时可能用到的各种 API，开发者通过这一层 API 构建自己的 App，基本的 API 如下：

(1) 视图系统：构建应用程序的界面。

(2) 内容提供者：允许应用程序访问其他应用的数据或共享数据。

(3) 通知管理器：允许应用程序在状态栏上显示定制的信息。

(4) 活动管理器：管理应用程序的生命周期。

(5) 资源管理器：提供对非代码资源的管理。

4. Android 应用层

应用层是一个核心应用程序的集合，所有安装在手机上的应用程序都属于这一层，例如短信、浏览器、通信录、微信、QQ、支付宝等。

1.3　Android Studio IDE 开发平台介绍

Android Studio 是基于 IntelliJ IDEA 的 Android 集成开发环境(IDE)，于 2013 年 5 月在谷歌 I/O 大会正式对外发布的，是一款为 Android 开发量身定制的编辑器。除了强大的代码编辑器和开发者工具，还提供了更多可提高 Android 应用开发效率的功能，主要特性如下：

(1) 基于 Gradle 的灵活构建系统。

(2) 快速且功能丰富的模拟器。

(3) 可针对所有 Android 设备进行开发的统一环境。

(4) 可将变更推送到正在运行的应用，无须构建新的 APK。

(5) 可帮助您构建常用应用功能和导入示例代码的代码模板和 GitHub 集成。

(6) 丰富的测试工具和框架。

(7) C ++和 NDK 支持。

(8) 内置对 Google 云端平台支持，可轻松集成 Google Cloud Messaging 和 App 引擎。

(9) 智能代码编辑器，能够非常高效的完成代码补全、代码重构和代码分析。

(10) 集成版本控制、代码分析工具和 UI 编辑器，Gradle 构建工具等强大的插件支持。

小　结

本章首先对目前主流的手机系统作了简单的介绍，包括苹果推出的 iOS 系统和谷歌的 Android 系统。随后介绍了 Android 系统的起源并简单地介绍了 Android 系统的特点。最后讲解了 Android 系统的体系结构。

课后练习

【微信扫码】
第 1 章课后练习

第 2 章

Android 开发环境

俗话说“工欲善其事，必先利其器”。进行软件开发，首要的工作是选择一个好的 IDE。优秀的 IDE 工具，可以大幅提高开发效率，节省大量的时间。以往的 Android 开发，都是采用 Eclipse+ADT 插件的 IDE 进行开发，但是随着 2013 年 Android Studio 的推出，该开发工具在 2015 年底已经被 Google 宣布停止维护，因此本书将只介绍基于 Android Studio 的 IDE 开发环境搭建。

2.1 Android Studio 开发环境的搭建

Android 的程序是基于 Java 语言的，因此首先需要 Java 环境，其次需要开发的工具包，最后需要集成的开发环境。包括如下 3 项：

（1）JDK：Java 语言的软件开发工具包，它包含了 Java 的运行环境、工具集合、基础类库等内容。本书中的 Android 程序必须要使用 JDK 1.8 或以上版本才能进行开发。

（2）Android SDK：谷歌提供的 Android 开发工具包，在开发中需要通过该工具包使用 Android 相关的 API。

（3）Android Studio：2013 年的时候，谷歌推出了一款官方的 IDE 工具——Android Studio。由于不再是以插件的形式存在，它在开发 Android 程序方面远比 Eclipse 强大和方便得多。

在 Windows 平台下，Android Studio 开发环境的搭建过程包括 5 个步骤：安装 JDK、安装 Android Studio、安装 Android SDK、配置 Android Studio、安装 Android 模拟器。

2.1.1 Java 开发环境的安装

【微信扫码】
Java 环境的安装

Android Studio 需要 JDK。下载的地址为 https://www.oracle.com/technetwork/java/javase/downloads/index.html。

下载页里面有 JDK 的好几个版本，可以根据实际情况选择下载，本书的代码至少需要 JDK 1.8 以上版本，同时还需要注意选择 32 位还是 64 位。JDK 安装完成之后，需要设置系统环境变量。假设 JDK 的安装路径为 C:\Java\jdk1.8.0_162_64bit，步骤如下：

（1）设置环境变量：右键单击计算机→属性→控制面板主页→高级系统设置→高级标签页→环境变量按钮→系统变量，新建变量名为“JAVA_HOME”，并将 JDK 安装路径作为变量值。

（2）编辑 Path：系统变量→Path→编辑，将“%JAVA_HOME%\bin;”添加到最前面。

（3）编辑 CLASSPATH：在系统变量中查找 CLASSPATH，如果没有则新建，变量名为“CLASSPATH”，变量值为“%JAVA_HOME%\lib”。

(4) 检验是否配置成功:运行 cmd 或者命令提示符,输入“java -version”,如果出现 Java 的版本信息,就表示安装成功了。

2.1.2 Android Studio IDE 的安装

Android Studio 的官方下载地址为 https://developer.android.google.cn/studio。下载页面打开如图 2-1 所示。

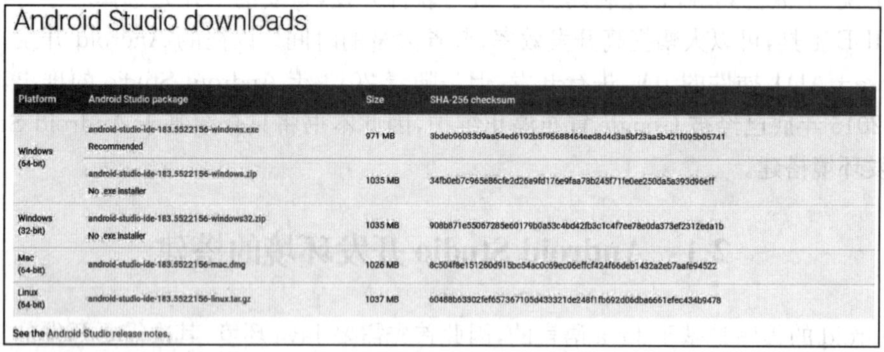

Android Studio downloads

Platform	Android Studio package	Size	SHA-256 checksum
Windows (64-bit)	android-studio-ide-183.5522156-windows.exe Recommended	971 MB	3bdeb96033d9aa54ed6192b5f95688464eed8d4d3a5bf23aa5b421f095b05741
	android-studio-ide-183.5522156-windows.zip No .exe installer	1035 MB	34fb0eb7c965e86cfe2d26a9fd176e9faa78b245f71fe0ee250da5a393d96eff
Windows (32-bit)	android-studio-ide-183.5522156-windows32.zip No .exe installer	1035 MB	908b871e55067285e60179b0a53c4bd42fb3c1c4f7ee78e0da373ef2312eda1b
Mac (64-bit)	android-studio-ide-183.5522156-mac.dmg	1026 MB	8c504f8e151260d915bc54ac0c69ec06effcf424f66deb1432a2eb7aafe94522
Linux (64-bit)	android-studio-ide-183.5522156-linux.tar.gz	1037 MB	60488b63302fef657367105d433321de248f1fb692d06dba6661efec434b9478

See the Android Studio release notes.

图 2-1 Android Studio IDE 下载页面

这里提供了 Windows、Linux、Mac 三大平台的安装包,Windows 平台还分为 32 位和 64 位两种安装包,用户根据自己计算机的操作系统选择对应的安装包进行下载。可以直接下载全部的安装包,下载好之后就可以进行安装,分为 6 个步骤。

步骤 1:运行 Android Studio 安装程序,点击 Next 进入 Choose Components 页面。

步骤 2:在组件选择界面中,显示四个组件需要安装:① Android Studio:必须安装。② Android SDK:安装 SDK,建议选择。③ Android Virtual Device:安装 Android 虚拟机,建议选择。④ Performance:安装 Intel 处理器的 HAXM 虚拟机加速服务,建议选择。这里建议全部勾选,然后点击 Next 按钮,进入许可证页面。

步骤 3:在许可证页面只能选择 I Agree,进入下一步 Configuration Settings 页面。

步骤 4:在 Configuration Settings 页面中确定安装目录。第一行选择 Android Studio 的安装目录,第二行选择 SDK 的安装路径,建议不要安装在有中文字符的目录之下。然后点击 Next 进入模拟器设置界面。

步骤 5:如果在 Choose Components 页面中选择了 HAXM 选项,就会出现设置模拟器的内存大小这一步,一般根据自己计算机的内存大小来确定,建议用默认即可。

【微信扫码】
Android Studio IDE 的安装

步骤 6:选择启动菜单中 Android Studio 程序显示的文件夹名称。然后点击 Install 开始安装,Android Studio 的运行需要 VC++,在安装中如果遇到禁止安装情况,请注意设置允许。等待安装,直到出现 Completing Android Studio Setup 界面,说明 Android Studio 已经安装成功了。接下来就是 Android SDK 的安装。

2.1.3　Android SDK 的安装

SDK 的全称是 Software Development Kit(软件开发工具包),它是开发程序必不可少的。在安装 Android Studio 之后,必须安装 SDK,否则不能创建项目。

安装 Android Studio 完成后,首次启动便会显示如图 2－2 所示界面,询问用户是否导入自定义配置文件。

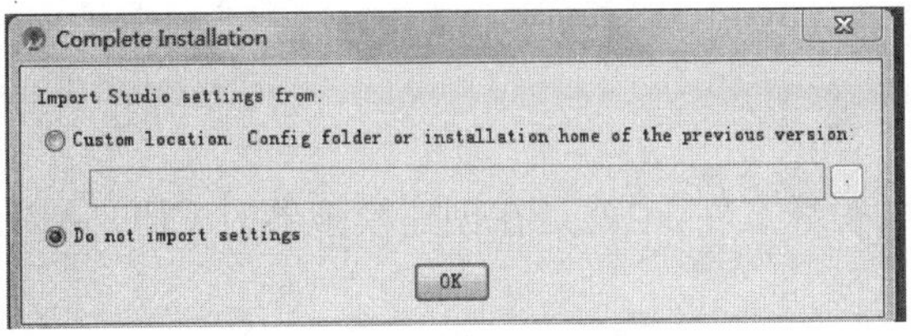

图 2－2　导入用户设置

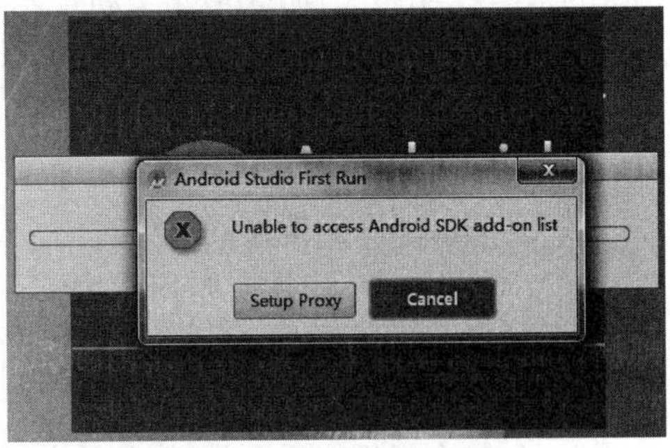

图 2－3　提示指定 SDK 路径

如果是首次安装或者不需要导入,则选择 Do not import settings,点击 OK,进入下一步。出现如图 2－3 所示对话框,表示检测不到 Android Studio 的 SDK,可能是安装的时候修改了 SDK 的默认安装目录,导致检测不到,可以稍后进行配置。直接点击 Cancel,出现欢迎界面之后,点击 Next,一直到 Finish,然后会出现下载 SDK 的界面。

正式下载 SDK 之前会执行 Fetching Android SDK component information,用于获取已安装的 SDK 组件信息,如图 2－4 所示,在这个界面,这个过程会等待比较长的时间,主要的原因是:网络连接有问题。可以采用修改 Hosts 的方法解决。配置完成之后进入下载和安装 SDK 的界面,如图 2－5 所示,这个过程耗时比较长,需要耐心等待,直到安装完成。

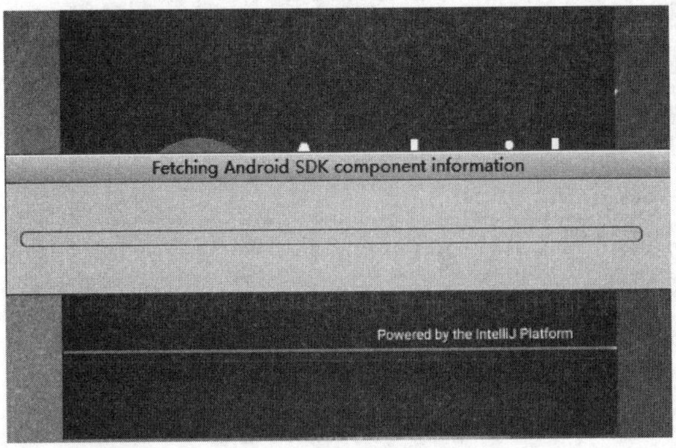

图 2－4　获得 SDK 组件的安装信息

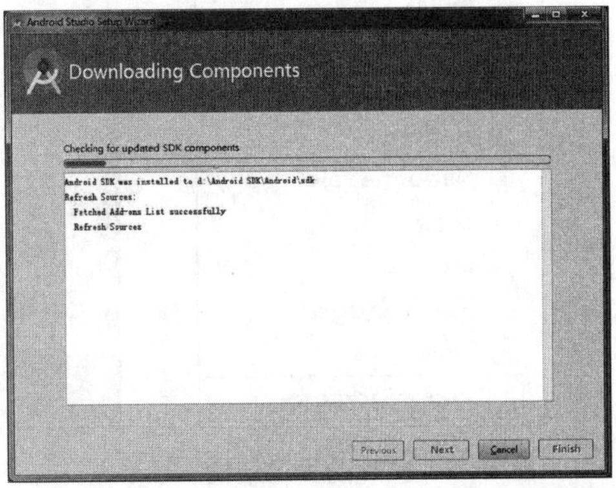

图 2－5　SDK 安装和更新过程

SDK 安装完成之后,打开安装路径查看 SDK 目录如图 2－6 所示,下面介绍下每个文件夹的作用。

build-tools:各版本 SDK 编译工具,构建项目时会用到,创建 Android 项目的时候也会用到这个包。在创建项目的时候如果没有此包会报错。

docs：离线开发者文档，Android SDK API 参考文档。所有的 API 都可以在这里查到。

emulators：Android SDK 模拟器主程序。

extras：扩展开发包，如高版本的 API 在低版本中开发使用的兼容包等。也会存放 Google 提供的 USB 驱动，Intel 提供的硬件加速附件工具包。

fonts：字体文件夹。

licenses：Android Market 版权保护组件。

patcher：增量更新，类似于更新记录。

platforms：平台的 SDK，里面根据 API Level 划分 SDK 版本。

platform-tools：Android 平台的相关工具，比如 adb.exe、sqlite3.exe 等以及一些通用工具，比如 adb、aapt、aidl、dx 等文件。

skins：Android 模拟器的皮肤。

sources：Android 各版本 SDK 源码。

system-images：创建 Android 模拟器时的镜像文件，也就是在创建模拟器时 CPU/ABI 项需要选择的。

tools：包含了很多重要的工具，比如用于启动 Android 调试工具的 DDMS。

2.1.4 Android Studio 的配置

安装 SDK 之后，就可以正常启动 Android Studio。启动之后在 Configure/Project Structure 对话框中，输入 JDK 和 SDK 的安装路径进行配置，如图 2－7 所示。NDK 暂时不需要配置。之后点击 OK 直至下一步，稍等片刻，Android Studio 开发环境就可配置成功。

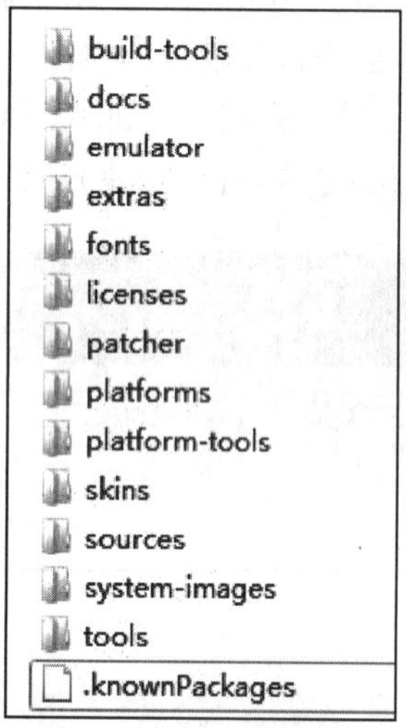

图 2－6　SDK 的目录结构

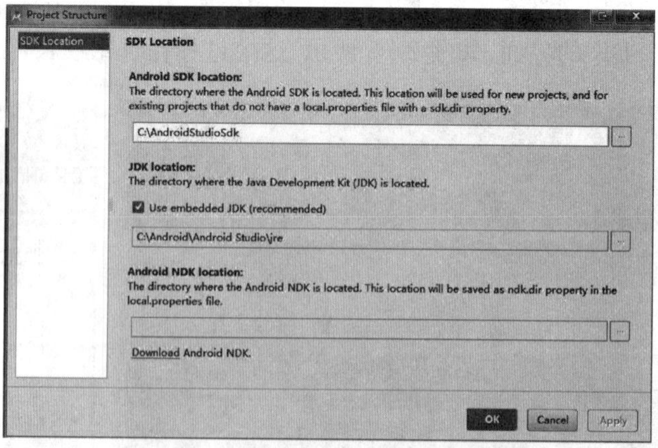

图 2－7　SDK 和 JDK 的安装路径

2.2 Android 模拟器的安装

运行 Android 程序有两种方法：第一种方法是用真实的安卓手机，但每次都把程序部署到手机上非常的麻烦；第二种方法是在电脑上使用 Android 模拟器，这样程序的调试和运行都比较方便。下面介绍模拟器的安装和使用。

(1) 启动模拟器管理器。在 Android Studio 中点击 Tools 菜单，选择 AVD Manager，出现模拟器管理界面。

(2) 创建模拟器：如果界面中没有模拟器，首先点击左下角的 Create Virtual Device 创建模拟器。接着在 Select Hardware 左侧的 Category 目录中，选择 Phone 并在中间的表格中确定模拟器的型号。同时右侧的窗口显示模拟器的外观和尺寸。点击 Next，进入镜像选择。

(3) 选择系统镜像：在 System Image 界面中，确定所使用的 API Level，一般情况下，选择版本号高的版本，在右边显示的是具体的信息。

(4) 编辑模拟器的信息：包括模拟器的名称、屏幕显示方向等。如果点击 Show Advanced Setting 按钮，可以进入模拟器高级设置，包括：Camera 设置前后摄像头的设置模式；Network 设置网络的访问模式；Emulated Performance 设置模拟器的图形加速模式、启动选项和 CPU 的数量等。Memory and Storage 设置存储器，Internal Storage 为内置存储器容量，SD card 为 SD 卡容量。Keyboard 中勾选 Enable keyboard input 允许模拟器在运行中，使用电脑键盘辅助输入，如果去掉勾选，则只能使用模拟器自带的键盘输入。设置好之后点击 Finish，完成设置。

模拟器创建完成之后，打开 C:\Users\用户\.android\avd 目录可以看到模拟器的文件，包括两个内容：.ini 后缀的是配置文件，.avd 后缀的是模拟器的文件夹。双击打开.avd 文件夹，文件夹中的内容显示如图 2-8 所示。这些文件是模拟器运行必备的，用户不能私自修改。模拟器创建完成后，点击▶启动，出现的模拟器如图 2-9 所示。

图 2-8　安卓模拟器.avd 文件夹

图 2-9　Android 模拟器

现在要对模拟器进行设置。在如图 2-9 所示的模拟器界面中，点击底部中间的小圆点，进入 Android 模拟器内置界面，如图 2-10 所示，再点击 Setting 进行模拟器的设置，如图 2-11 所示。

图 2 - 10　模拟器内部界面

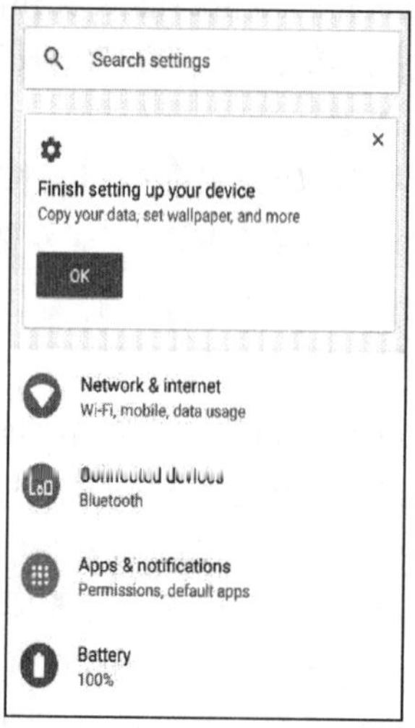

图 2 - 11　模拟器 Setting 界面

设置内容很多，这里只通过语言环境来介绍模拟器的设置。

在模拟器 Setting 界面向下滑动，拉到最下面找到 System，点击进入 System 后点击 Languages & input 设置语言和输入法。直接点击进行语言环境设置，系统预设的只有英语，点击 Add a language，添加语言，向下滑动到底，找到“简体中文”。在简体中文的区域中直接选择“中国”，跳转回到语言属性界面，这时语言设置有原来的英语和新添加的中文。点击左上角的菜单图标，弹出 Remove 后选择删除 English，如图 2 - 12 所示。当看到菜单条上原来的 Language preferences 变为了中文的“语言偏好设置”，如图 2 - 13 所示，说明中文语言设置成功。

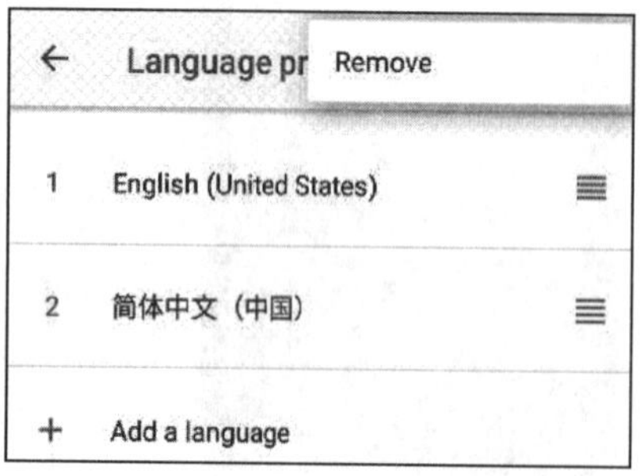

图 2 - 12　删除英文

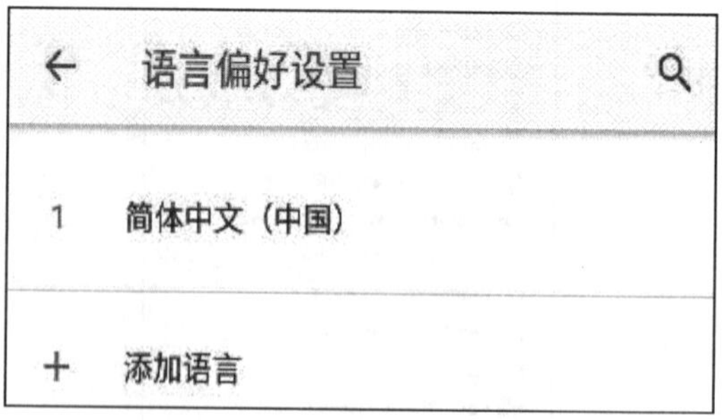

图 2 - 13　简体中文添加完成

2.3　Android 项目初体验

2.3.1　项目的创建和启动

安装好开发环境就可以创建 Android 项目了。点击 Android Studio 的图标启动应用程序。出现 Welcome to Android Studio 界面，按照下面的步骤创建第一个项目。

步骤 1：点击 Start a new Android Studio project 创建第一个项目。

步骤 2：选择 Activity 模板界面，在这个界面，选择 Activity 的模板。Google 提供了一些带有基础组件的 Activity，方便开发者的使用。作为初学者，我们选择 Empty Activity。然后点击 Next 进入下一步。

步骤 3：设置项目的基本信息。Name 为项目的名称，Package name 为项目程序包的名称，Save location 确定项目存放的路径，存放的路径不要有中文或者空格。Language 是开发语言，Minimum API level 是项目使用的 API。初学者不用修改，直接使用默认配置信息即可。点击 Finish，耐心等待一会，第一个 Android 项目就创建成功了。

步骤 4：在 Android Studio 的 Run 菜单中，点击 Run 'app'，开始对项目代码进行编译，编译通过之后会启动已配置好的模拟器，当模拟器界面上出现 Hello World 字样时，表示这个项目已经安装成功。

上面的步骤中，开发者没有写一行代码，项目就运行起来了。这归功于 Android Studio 非常人性化，一些简单内容都能自动生成。下面我们先分析一下 Android 项目的结构。

2.3.2　Android 项目的目录结构

使用 Project 模式查看，可以看到 Android 项目的目录结构，如图 2－14 所示。对结构中的文件夹和文件进行介绍。

（1）.gradle 和.idea：这两个目录下放置的都是一些自动生成的文件，不要手动编辑。

（2）app：项目中的代码、资源等内容几乎都是放置在这个目录下的，后面的开发工作也基本都是在这个目录下进行的，后面还会对这个目录单独展开进行讲解。

（3）build：主要包含了一些在编译时自动生成的文件，不要去手动编辑。

（4）gradle：这个目录下包含了 gradle wrApper 的配置文件，使用 gradle wrApper 的方式不需要提前将 gradle 下载好，而会自动根据本地的缓存情况决定是否需要联网下载 gradle。Android Studio 默认没有启动 gradle wrApper 的方式，如果需要打开，可以点击 Android Studio 导航栏→File→Settings→Build，Execution，Deployment→Gradle，进行配置更改。

（5）.gitignore：这个文件是用来将指定的目录或文件排除在版本控制之外的。

（6）build.gradle：这是项目全局的 gradle 构建脚本，这个文件中的内容是不需要修改的。

（7）gradle.properties：这个文件是全局的 gradle 配置文件。

（8）gradlew 和 gradlew.bat：这两个文件是用来在命令行界面中执行 gradle 命令的，其中 gradlew 是在 Linux 或 Mac 系统中使用的，gradlew.bat 是在 Windows 系统中使用的。

（9）HelloWorld.iml：iml 文件是所有 IntelliJ IDEA 项目都会自动生成的一个文件。

（10）local.properties：这个文件用于指定本机中的 Android SDK 路径，通常内容都是自动生成的。如果 Android SDK 位置发生了变化，将这个路径改成新的位置即可。

（11）settings.gradle：用于指定项目中所有引入的模块，引入自动完成不需要手动修改。

图 2-14 Android 的工程目录

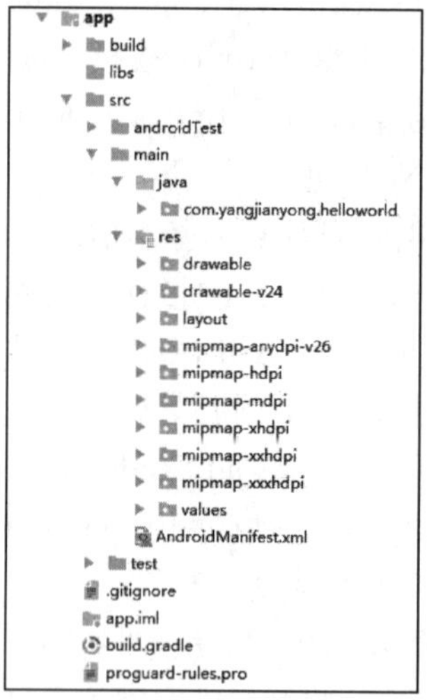

图 2-15 app 目录结构

如果切换成 Android 模式,将显示为 app 目录结构,如图 2-15 所示。

(1) libs:如果使用了第三方 jar 包,需要把这些 jar 包都放在 libs 目录下,这些 jar 包都会被自动添加到构建路径中。

(2) androidTest:用来编写 Android Test 测试用例。

(3) java:放置所有 Java 代码的地方。

(4) res:项目中使用到的所有图片、布局、字符串等资源都要存放在这个目录下。

(5) test:此处是用来编写 Unit Test 测试用例的,是对项目进行自动化测试的一种方式。

(6) app.iml:IntelliJ IDEA 项目自动生成的文件。

(7) build.gradle:这是 App 模块 gradle 构建脚本,会指定项目构建相关配置。

(8) proguard-rules.pro:这个文件用于指定项目代码的混淆规则,当代码开发完成后打包成安装包文件,如果不希望代码被别人破解,通常会将代码混淆,从而让破解者难以阅读。

2.3.3 项目的资源文件

在 app 目录中,res 目录下面存放的是项目的资源文件,里面的文件夹和文件很多,如图 2-16 所示,下面进行详细的介绍。

以 drawable 和 mipmap 开头的文件夹是用来放图片的。mipmap 简称为纹理映射技术,支持对 bitmap 图片的渲染,提高渲染的速度和质量,主要对图片大小缩放进行处理,降低图片的失真率。建议大家把 App 启动图标放在 mipmap 目录中,应用程序使用的图片资源放在 drawable 或 mipmap 目录中。以 mipmap 开头的多个文件夹,主要给一张图片提供几个不同分辨率的版本,让不同分辨率的设备加载合适分辨率的图片,建议的尺寸如下:mipmap-mdpi(48×48)、mipmap-hdpi(72×72)、mipmap-xdpi(96×96)、mipmap-xxdpi(144×144)、mipmap-xxxdpi(192×192)。如果只有一张图片,将图片放入 mipmap-xxdpi 文件夹即可。

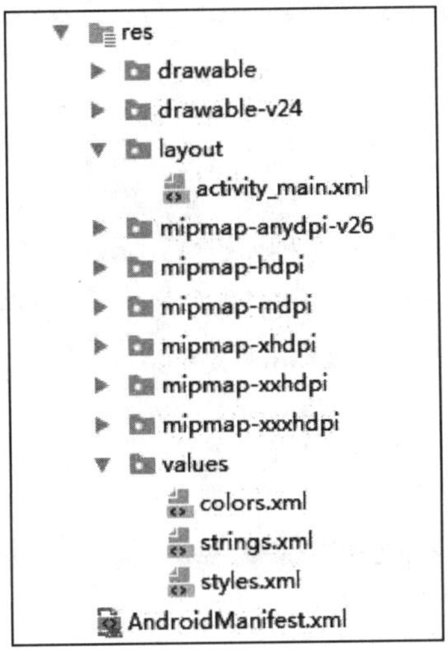

图 2－16　Android 的资源文件夹

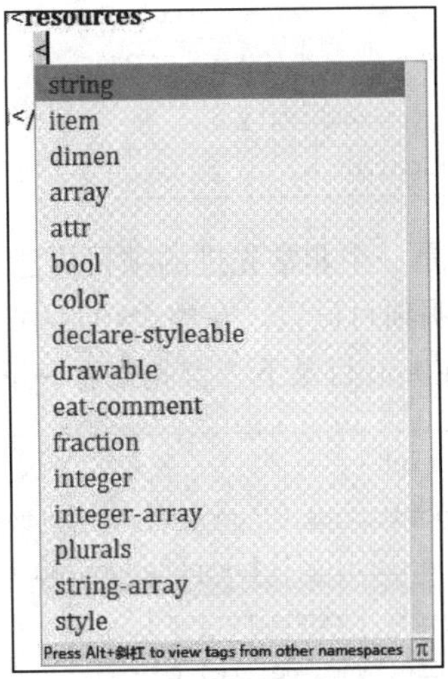

图 2－17　Android 所支持的资源类型

layout 文件夹用来存放布局文件。项目创建时会自动新建 activity_main.xml 文件。

values 文件夹用来存放资源文件，并按照资源类型分类，每个资源的 XML 文件都使用< resources ></resources >作为资源根标签，不同类型资源的子标签各有不同。

(1) strings.xml 为字符串资源，子标签为< string name="XXXXXX">……</ string >。

(2) colors.xml 为颜色资源，子标签为< color name="XXXXXX ">……</ color >。

(3) styles.xml 为样式资源，子标签分为两级。

```
< style name = "AppTheme" parent = "XXXXXX ">
        < item name = " XXXXXX">……< /item >
< /style >
```

其他类型资源这里不再详述。

2.3.4　项目的配置文件

AndroidManifest.xml：是整个 Android 项目的配置文件，程序中定义的组件都需要在这个文件里注册，另外还可以在这个文件中给应用程序添加权限声明。主要的代码如下。

```
< Application android:allowBackup = "true"
        android: icon = "@mipmap /ic_launcher" //指定应用程序的图标
        android:label = "@string /App_name" //指定应用程序的名称
        android:roundIcon = "@mipmap /ic_launcher_round"
        android:supportsRtl = "true"   android:theme = "@style /AppTheme">
        < activity android:name = ".MainActivity"> //应用程序启动界面
            < intent-filter >
                < action android:name = "android.intent.action.MAIN"  />
```

```
            <category android:name = "android. intent. category. LAUNCHER"  />
        </intent-filter>
    </activity>
</Application>
```

Gradle 是一个非常先进的项目构建工具。它使用了一种基于 Groovy 的领域特定语言(DSL)来声明项目设置。在整个项目中有两个 build.gradle 这样的文件,一个在最外层目录下,另一个在 app 目录下。首先来看一下最外层的 build.gradle 文件。

```
buildscript {
    repositories {   google()   jcenter()    }
    dependencies {classpath 'com. android. tools: build:gradle:3. 0. 0' }}
allprojects { repositories {   google()    jcenter()   }}
task clean(type: Delete) {……}
```

jcenter()这个配置主要是使用代码托管仓库。如果将代码托管到 jcenter 上,就可以轻松引用上面的任何开源项目。dependencies 中使用 classpath 声明了一个 gradle 插件,最后面字符串是插件的版本号。这些代码都是自动生成的,不需要手动修改其中的内容。

app 目录下面的另一个 build.gradle 文件。这个文件比较复杂,主要包含如下内容:

apply plugin 表明这里应用了一个插件,是应用程序模块。

android 的大闭包中是项目配置的属性。compileSdkVersion 表示项目编译的版本,applicationId 用于指定项目的包名,minSdkVersion 用于指定项目最低兼容的 Android 系统版本,targetSdkVersion 用于指定项目运行的 Android 系统版本,versionCode 表示项目版本号,versionName 表示项目版本名。另外在项目编译时,还会加上 buildToolsVersion,确保构建工具的版本一致。上述用到的版本必须在 SDK 中有。

buildTypes 闭包用于指定生成安装文件的相关配置。release 用于指定生成正式安装文件的配置。其中的内容 minifyEnabled 用于指定是否对项目代码进行混淆,proguardFiles 用于指定混淆时用的规则文件。混淆的作用是保护代码,不会被别人轻易破解。

dependencies 闭包是指明项目依赖库,这个包的功能在项目运行中是非常重要的。

implementation fileTree(dir: 'libs', include: ['*.jar']),表示将本地 libs 目录下的所有.jar 后缀的文件都添加到项目的构建路径中。例如 implementation 'com. android. support:Appcompat-v7:26.0.0-beta1' 表示加载远程依赖库,在 Gradle 构建的时候,首先检查一下本地是否已经有了这个库的缓存,如果没有则去网上下载,添加到本地的依赖库中。

如果有其他第三方依赖库,用如下的方式添加进来即可:Implementation '******', ****** 为第三方依赖库的名称。

testImplementation 和 androidTestImplementation 用于声明测试用例,暂时用不到。

【微信扫码】
build.gradle
详细内容

2.4　Android Studio 的使用

2.4.1　Android Studio 的界面

打开 Android Studio IDE，界面如图 2－18 所示，下面对其做一些介绍。

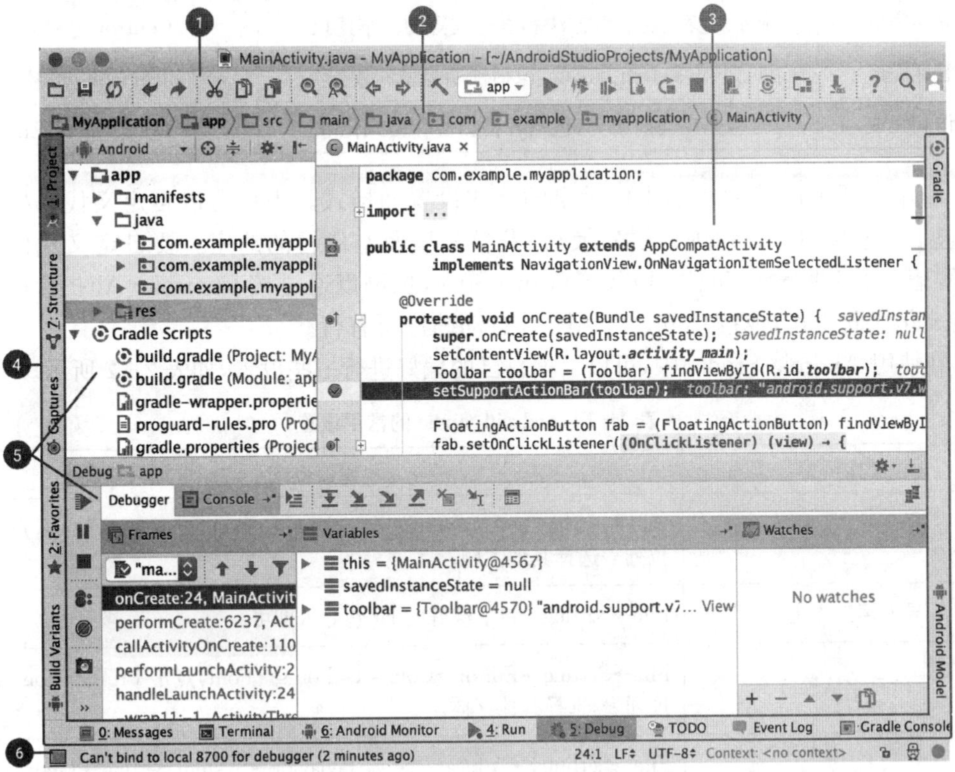

图 2－18　Android Studio 的界面

❶ 工具栏：提供执行各种操作的工具，包括运行应用和启动 Android 工具。

❷ 导航栏：帮助用户在项目中导航，以及打开文件进行编辑。

❸ 编辑器：窗口是创建和修改代码的区域。

❹ 工具窗口栏：在 IDE 窗口外部运行，并且包含可用于展开或折叠各个工具窗口的按钮。

❺ 工具窗口：提供对特定任务的访问，例如项目管理、搜索和版本控制等。

❻ 状态栏：显示项目和 IDE 本身的状态以及任何警告或消息。

2.4.2　常用命令

常用的工具窗口固定在应用窗口边缘的工具窗口栏上。要展开或折叠工具窗口，请在工具窗口栏中点击该工具的名称。还可以拖动、固定、取消固定、关联和分离工具窗口。要返回到当前默认工具窗口布局，请点击 Window→Restore Default Layout 或点击

Window→Store Current Layout as Default 自定义默认布局。

在编写代码过程中，写出正确的类名、方法等是比较烦琐的。这时候，代码自动完成的功能就非常有用，能在用户输入字符串的部分内容时，提供下拉菜单，自动推荐相关常用字符串供用户选择。Android Studio 提供的代码自动完成功能如表 2-1 所示。

表 2-1　Android Studio 代码自动完成

类型	说明	操作方式
基本完成	显示对变量、类型、方法和表达式等的基本建议。	Control＋空格
智能完成	由上下文显示相关选项，识别预期类型和数据流。	Control＋Shift＋空格
语句完成	自动完成当前语句，包括各种括号和格式化等。	Control＋Shift＋Enter

Android Studio 会自动匹配已设定的格式设置和样式。用户可以自定义代码样式设置，包括指定选项卡和缩进、空格、换行、花括号以及空白行的约定。要自定义代码样式设置，请点击 File→Settings→Editor→Code Style，也可以通过按 Control＋Alt＋L 调用重新格式化代码操作，或按 Control＋Alt＋I 自动缩进所有行。

在使用 Android Studio 中，可以根据自己的偏好进行一些设置，如表 2-2 所示。

表 2-2　Android Studio 的常用设置

操作类型	操作方式
主题修改	File→Settings→Appearance & Behavior→Appearance→Theme，点击下拉列表选择其他主题。
导入第三方主题	File→Settings，将下载好的 jar 包导入即可。
代码字体修改	File→Setting→Editor→Colors & Fonts→Font，点击 Primary font 的下拉列表，选择其他字体。
关闭更新	File→Settings→Appearance & Behavior→System Setting→Updates，去掉 Automatically check updates for 的勾选状态。
快捷键习惯的修改	File→Settings→Keymap，在下拉列表中选择。
配置代码自动提示	File→Settings→Editor→General→Code Completion，将 Autopopup code completion 红框部分处于勾选状态。
禁止代码折叠	File→Settings→Editor→General→Code Folding 中将红框标出的设置为去掉勾选。

2.4.3　日志输出和调试

开发程序时，掌握日志的输出是非常必要的。学习 Android 必须掌握这个知识点。日志可以借助 Android Monitor 工具来查看，如图 2-19 所示，默认的快捷键是 Alt＋6。

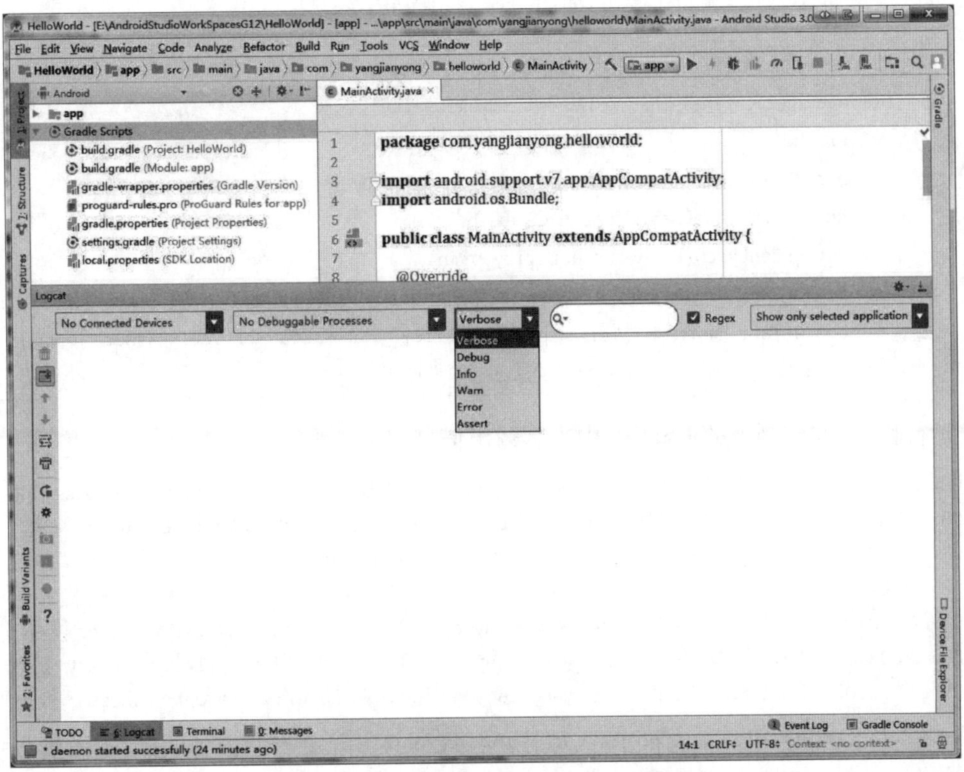

图 2-19　Android Studio 的 LogCat 界面

从图 2-19 可以看到，LogCat 提供了六级的日志输出模式，包括 Verbose、Debug、Info、Warn、Error、Assert，级别依次升高，见表 2-3。每条日志都有进程信息，默认一个应用程序占一个进程，且每个日志都有标签。

表 2-3　系统日志说明

模式	级别	描述	Log
所有	Verbose	不加过滤的输出所有的调试信息。包括 Verbose、Debug、Info、Warn、Error、Assert 级别。	Log.v()
调试	Debug	输出调试信息中的基础信息。包括 Debug、Info、Warn、Error 级别。	Log.d()
信息	Info	用于打印一些比较重要的数据，可以帮你分析。包括 Info、Warn、Error 级别。	Log.i()
警告	Warn	用于打印一些警告信息，提示程序在这个地方可能会有潜在的风险。输出包括 Warn、Error 级别。	Log.w()
错误	Error	用于打印程序中的错误信息，一般都代表程序出现严重问题了，必须进行修复。输出包括 Error 级别。	Log.e()
断言	Assert	当程序中的断言表达式失败，则抛出异常信息。	Log.wtf()

可以写一个小程序,验证日志的输出方法,代码如下。

```
public class MainActivity extends AppCompatActivity {
    private static final String TAG = "MainActivity";
    @Override
    protected void onCreate(Bundle savedInstanceState) {
        super.onCreate(savedInstanceState);
        setContentView(R.layout.activity_main);
        Log.v(TAG, "verbose 模式");Log.d(TAG, "debug 模式");Log.i(TAG, "info 模式");
        Log.w(TAG, "warn 模式");Log.e(TAG, "error 模式");Log.wtf(TAG, "assert 模式");
        System.out.println("system 的输出模式"); }}
```

运行程序,在控制台的 LogCat 中查看输出信息如下。

```
08-28 02:03:57 325 6219-6219 /com.yangjianyong.helloworld V /MainActivity: verbose 模式
08-28 02:03:57 325 6219-6219 /com.yangjianyong.helloworld D /MainActivity: debug 模式
08-28 02:03:57 325 6219-6219 /com.yangjianyong.helloworld I /MainActivity: info 模式
08-28 02:03:57 325 6219-6219 /com.yangjianyong.helloworld W /MainActivity: warn 模式
08-28 02:03:57 325 6219-6219 /com.yangjianyong.helloworld E /MainActivity: error 模式
08-28 02:03:57 325 6219-6219 /com.yangjianyong.helloworld E /MainActivity: assert 模式
08-28 02:03:57 325 6219-6219 /com.yangjianyong.helloworld I /System.out: system 的输出模式
```

在编码过程中会出现很多的 Bug,要找出这样的 Bug 是非常耗时耗力的,这就需要程序员熟练掌握各种调试工具的使用,快速准确地定位 Bug 所在的位置和出现的原因。

Android Studio 提供了 DDMS 工具用于程序的调试监控服务。DDMS 全称为 Dalvik Debug Monitor Service。在 Android Studio 中,没有将 DDMS 的工具放置在工具栏上,可以通过如下方式打开 DDMS 窗口:在 SDK 的安装目录中,找到 tools 文件夹,点击里面的 monitor.bat 就可以打开 DDMS 窗口,如图 2-20 所示。

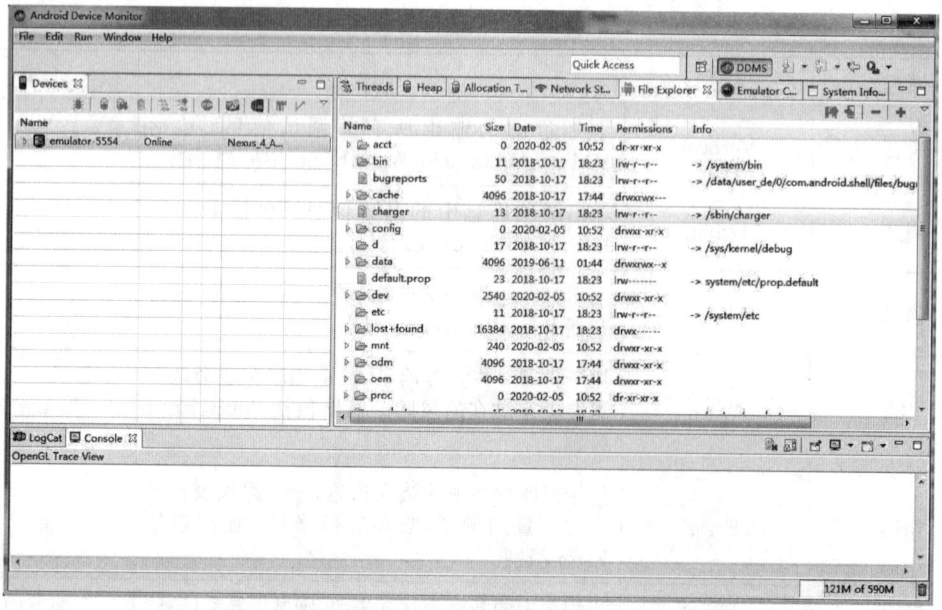

图 2-20　DDMS 界面

在 DDMS 中最重要的两个功能分别是 Device(设备列表)和 File Explorer(文件列表)。Device 是设备的列表窗口,可以查看在运行中的所有模拟器和各个模拟器中运行的所有进程。在显示列表的最上面,是当前正在运行的模拟器的名称。每个模拟器项目条下面列表的第一列,是 Android 应用程序的进程名,按照对应的包名进行组织,中间一列是应用程序的进程号 PID,最右边一列是应用程序的端口号。File Explorer 文件浏览器是 Android 的文件系统,类似于 Linux 的文件系统,可以很方便地在模拟设备和计算机之间导入和导出文件。

小　结

本章重点介绍了如何搭建 Android 开发环境,包括环境搭建、熟悉环境、创建项目的步骤、项目结构、调试程序、日志输出等。本章内容是进行 Android 开发的必备知识,读者必须熟练掌握。

课后练习

【微信扫码】
第 2 章课后练习

第 3 章

界面设计和资源管理

Android 系统开发的第一项工作，就是用户界面的开发。一个应用程序不管包含多少复杂的逻辑程序，如果没有提供友好的用户图形界面，最终无法吸引用户。

Android 提供了大量功能丰富的 UI 组件，开发者只需要像“搭积木”一样，将这些 UI 组件组合在一起，就可以开发出优秀的图形用户界面，并通过事件响应机制，让这些组件能响应用户的鼠标、键盘动作。

3.1 界面设计

用户运行程序，首先看到的就是程序的界面。用户对应用程序的评价很大程度上取决于界面的好坏。因此任何应用程序的设计都离不开优良的界面设计。Android 的界面非常出色，不光可以通过 XML 直接设计，还可以通过 Java 编程方式实现，对于复杂的界面，还可以通过 XML+Java 结合的方式来实现，下面分别介绍这三种方式。

3.1.1 XML 方式

Android 推荐使用 XML 布局文件来实现界面，不仅简单明了，而且可以将应用的视图控制逻辑从 Java 代码中分离出来，放入 XML 文件中，从而更好地体现 MVC 原则。

下面通过一个案例来介绍 XML 实现界面的方法。

案例 3 - 1：使用 XML 配置布局文件

【微信扫码】
案例 3 - 1
相关文件

新建一个 Android 项目，只使用 XML 来设计界面，命名为 Basic_Teach_XMLLayout。

步骤 1：编写字符串资源文件

在 app/res/values/strings.xml 文件的< resources ></resources >标签内，添加代码如下。

```
< string name = "App_name">欢迎使用安卓程序< /string >……
```

步骤 2：编写布局文件

在 app/res/layout/activity_main.xml 布局文件中，使用垂直方向的线性布局作为容器，在其中放置 2 个文本框，第 1 个靠左放置，第 2 个靠右放置，按钮居中放置。主要代码如下。

```
< LinearLayout android:orientation = "vertical" >
    < TextView android: layout _ width = " wrap _ content" android: layout _ height = " wrap _
content" android:text = "@string /hello_world" />
    < TextView android:layout_gravity = "right" android:text = "@string /txetview" />
    < Button android:layout_gravity = "center_horizontal"  />
< /LinearLayout >
```

编写完成后，不需要修改 MainActivity 的任何代码，直接运行程序，效果如图 3－1 所示。

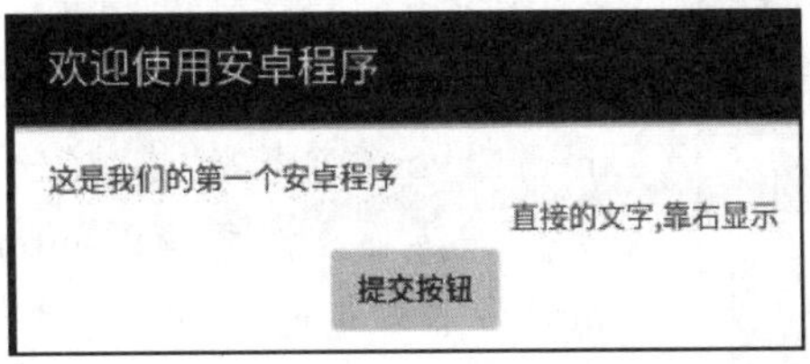

图 3－1　使用 XML 布局文件实现界面布局效果

3.1.2　Java 控制

虽然 Android 推荐使用 XML 实现 UI 界面，但 Android 也允许开发者像开发 Swing 应用一样，完全抛弃 XML 布局文件，在 Java 代码中实现 UI 界面。如果在代码中实现 UI 界面，那么所有的 UI 组件都将通过关键字 new 创建出来，然后以合适的方式“搭建”在一起即可。下面我们完全使用 Java 进行界面编程。

【微信扫码】案例 3－2 相关文件

案例 3－2：使用 Java 代码来生成控件

新建一个 Android 项目，命名为 Basic_Teach_JavaLayout，在本案例中，没有编写 XML 的布局文件，直接修改 src/MainActivity.java 中的代码如下。

```
public class MainActivity extends Activity {
    private int[]  colorarray = new int[]{Color. BLUE, Color. GREEN, Color. RED, Color. DKGRAY,
Color. GRAY, Color. LTGRAY}; //定义颜色资源数组
    @Override
    protected void onCreate(Bundle savedInstanceState) {
      super. onCreate(savedInstanceState);
      LinearLayout linearlayout = new LinearLayout(this);  //生成线性布局
      LinearLayout. LayoutParams linearParams = new LinearLayout. LayoutParams(ViewGroup.
LayoutParams. MATCH_PARENT, ViewGroup. LayoutParams. MATCH_PARENT); //设置布局的尺寸参数
            linearlayout. setOrientation(LinearLayout. VERTICAL);  //设置垂直对齐
            linearParams. gravity =  Gravity. CENTER_VERTICAL;  //设置组件居中对齐
            linearlayout. setLayoutParams(linearParams); //参数加载到线性布局中
            super. setContentView(linearlayout); //线性布局加载到布局中
            for (int i = 0;i<6;i ++ ){ //使用循环方式生成 6 个按钮
              Button bnt = new Button(this);
```

```
                bnt.setText("使用 java 代码生成的按钮" + i); //设置文本
                bnt.setBackgroundColor(colorarray[i]); //设置按钮的颜色
                    LinearLayout.LayoutParams bntparams = new LinearLayout.LayoutParams
(ViewGroup.LayoutParams.WRAP_CONTENT,ViewGroup.LayoutParams.WRAP_CONTENT);
                bntparams.gravity = Gravity.CENTER; //按钮的对齐方式
                bnt.setLayoutParams(bntparams); //将按钮的设置参数加载到按钮中
                linearlayout.addView(bnt); }  }}
```

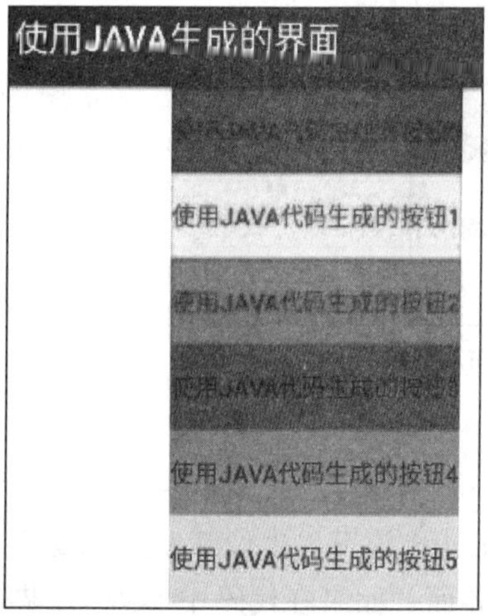

图 3-2　使用 Java 实现界面布局的效果

运行程序，效果如图 3-2 所示。

从上面的程序可以看出，该程序中所用到的 UI 组件都是通过 new 关键字创建出来的，然后使用 LinearLayout 容器来“盛装”这些 UI 组件，这样就组成了图形用户界面。注意 setContentView(R.layout.activity_main)这个语句是文件创建时自动生成的，表示调用 res/layout 中的布局文件。因为这个程序是使用 Java 来创建布局，不需要使用布局文件，所以将这个语句注释掉。这里需要提醒的是，在使用 findViewById()方法之前，必须先使用 setContentView 方法装载 XML 布局文件，否则系统会抛出异常。

3.1.3　XML＋Java 方式

从上面的两个例子中可以看出，完全使用 XML 文件实现 UI 界面虽然方便，但难免有失灵活。完全使用 Java 代码实现 UI 界面不仅烦琐，而且不利于解耦。因此有时候，需要将 XML 布局文件和 Java 代码混合使用来实现 UI 界面。把变化小、比较固定的组件放在 XML 文件中，而那些变化较多、行为控制比较复杂的组件交给 Java 代码来实现。布局文件中添加多个 UI 组件，可以为该 UI 组件指定 android:id 属性，Java 代码中可通过 findViewById(R.id.属性的 ID 值)代码来访问它。

一旦在程序中获得指定 UI 组件之后，接下来就可以通过代码来控制各 UI 组件的外观行为，包括为 UI 组件绑定事件监听器等。下面我们使用 XML＋Java 实现界面和控件。

【微信扫码】
案例 3-3
相关文件

案例 3-3：使用 XML 和 Java 配置布局文件

新建一个 Android 项目，命名为 Basic_Teach_XMLJavaLayout。

步骤 1：打开 app/res/layout/activity_main.xml 文件编写布局文件，修改代码如下。

```
<LinearLayout android:layout_width = "match_parent"   android:layout_height = "match_parent"
android:orientation = "horizontal"   android:id = "@ + id/layout" ></LinearLayout>
```

布局文件中简单放置一个线性布局文件，其他组件通过 Java 代码实现。

步骤 2：打开 app/src/MainActivity.java 编写实现代码，修改代码如下。

```
public class MainActivity extends Activity {
    private int[] imagePath = new int[]{R.drawable.img01,R.drawable.img02,R.drawable.
img03}; //整形数组方式获得图片
    private ImageView[] img = new ImageView[3]; //定义显示图片的组件
    @Override
   protected void onCreate(Bundle savedInstanceState) {
        super.onCreate(savedInstanceState);setContentView(R.layout.activity_main);
        LinearLayout layout = (LinearLayout)findViewById(R.id.layout); //加载线性布局
   for (int i = 0;i<imagePath.length;i++){ //用循环的方式来加载图片
        img[i] = new ImageView(this); //获得图片显示的组件
        img[i].setImageResource(imagePath[i]); //将对应的图片加载在组件上
        img[i].setPadding(3, 3, 3, 3); //组件和周围的边距
        ViewGroup.LayoutParams params = new ViewGroup.LayoutParams(240, 140);
        img[i].setLayoutParams(params); //将设置好的参数加载到控件上
         layout.addView(img[i]); //组件加载到布局文件    }  }  }
```

运行程序,效果如图 3－3 所示。

图 3－3　使用 XML＋Java 实现界面布局的效果

3.2　基本布局

Android 系统通过布局管理器,管理 UI 界面中的各个组件。通常来说,推荐使用布局管理器来管理组件的大小,而不是在程序中直接设置组件的位置和大小。针对不同的手机屏幕分辨率,布局管理器还可以根据运行的平台来调整组件的大小。Android 的布局管理器本身就是一个 UI 组件,都是 ViewGroup 的子类。而 ViewGroup 又继承于 View,因此,布局管理器还可以作为普通的 UI 组件使用。

Android 的基本布局有线性、表格、相对、帧和百分比布局等,线性布局在界面设计的三个案例中已经使用了,接下来将介绍其他几个布局。

3.2.1　表格布局

表格布局由 TableLayout 代表,TableLayout 继承于 LinearLayout,因此它完全可以支持 LinearLayout 的全部 XML 属性。向 TableLayout 中添加组件,该组件将直接占用一行。列的宽度由该列中最宽的那个组件决定。它在本质上依然是线性布局。采用行、列方式来管理 UI 组件。在使用时,不需首先确定多少行、多少列,而是通过添加 TableRow,

并在其中添加组件来控制布局的行列数。表格布局管理器常用的 XML 属性如表 3 - 1 所示。

表 3 - 1　表格布局的 XML 属性

控件属性	功能描述
android:shrinkColumns	如果值域添加了某个列,那么该列的所有单元格的宽度可以收缩,确保表格能适应父容器的宽度。
android:stretchColumns	如果值域添加了某个列,那么该列的所有单元格的宽度可以拉伸,确保表格能适应父容器的宽度。
android:collapseColumns	如果值域添加了某个列,那么该列的所有单元格会被隐藏。

下面通过一个案例来介绍表格布局的使用。

案例 3 - 4:使用表格布局实现一个登录界面的布局

本案例将完成一个登录界面,通过表格布局来实现对控件的布置。界面上有输入框和登录按钮等。新建一个项目,命名为 Layout_Teach_TableLayout,打开 res/layout 文件夹,修改布局文件 activity_main.xml 文件,主要代码结构如下。

【微信扫码】
案例 3 - 4
相关文件

```
< TableLayout android:gravity = "center_horizontal"……>
        < TableRow android: id = "@ + id /tablerow1"……>
                < TextView android: id = "@ + id /textview1"…… />
                 < EditText android: id = "@ + id /edittext1"…… />
        < /TableRow >
        < TableRow android: id = "@ + id /tablerow2"……>
            < TextView android: id = "@ + id /textview2"…… />
           < EditText android: id = "@ + id /edittext2" android: inputType = "textPassword"…… />
        < /TableRow >
          < TableRow android: id = "@ + id /tablerow3"……>
              < Button android: id = "@ + id /bnt1" android: text = "登录"…… />
              < Button android: id = "@ + id /bnt2" android: text = "退出"…… />
        < /TableRow >
  < /TableLayout >
```

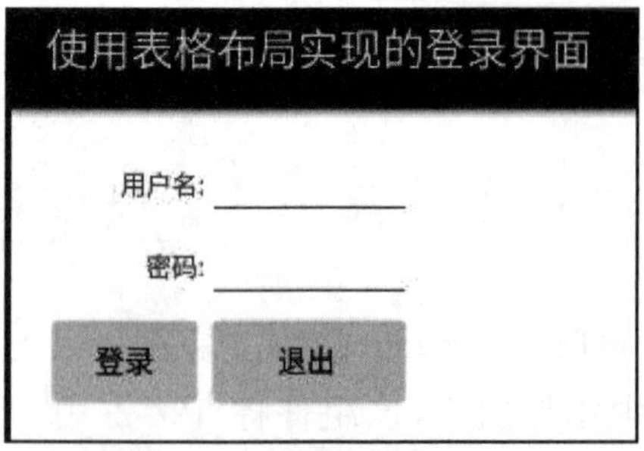

图 3 - 4　使用表格布局实现登录的界面效果

在代码中,TableLayout 标签下通过 TableRow 将表格分为三行,每行中再布置控件。表格布局的每行内没有列的概念,直接通过控件来区分。因为不实现真正的登录功能,不需要修改 Activity 代码直接运行,效果如图 3 - 4 所示。

3.2.2　相对布局

相对布局是通过相对定位的方式指定控件位

置,即以其他控件或父容器为参照物,摆放控件位置。设计相对布局时,要注意控件之间的位置依赖关系,后放入控件的位置依赖于先放入的控件。相对布局的属性较多,表 3 - 2 介绍了比较重要的位置属性。

表 3 - 2　相对布局的位置属性

控件属性	功能描述
android:layout_alignParentTop	设置当前控件是否与父控件顶端对齐。
android:layout_alignParentLeft	设置当前控件是否与父控件左边对齐。
android:layout_alignParentRight	设置当前控件是否与父控件右边对齐。
android:layout_alignParentBottom	设置当前控件是否与父控件底部对齐。
android:layout_alignParentStart	设置当前控件起始边缘是否与父控件起始边缘对齐。
android:layout_alignParentEnd	设置当前控件末端边缘是否与父控件末端边缘对齐。
android:layout_above	设置当前控件位于某控件上方。
android:layout_below	设置当前控件位于某控件下方。
android:layout_alignTop	设置当前控件的上边界和某控件的上边界对齐。
android:layout_alignLeft	设置当前控件的左边界和某控件的左边界对齐。
android:layout_alignRight	设置当前控件的右边界和某控件的右边界对齐。
android:layout_alignBottom	设置当前控件的下边界和某控件的下边界对齐。

下面我们使用相对布局来实现一个图片的分布效果。

案例 3 - 5:相对布局的使用

新建一个 Android 项目,命名为 Layout_RelativeLayout。打开 res/layout/activity_main.xml 文件,修改布局代码如下。

【微信扫码】
案例 3 - 5
相关文件

```
< ImageView android:id = "@ + id /imagecenter" android:src = "@drawable /center"
        android:layout_centerInParent = "true"………… />
< ImageView android:id = "@ + id /imagephone" android:src = "@drawable /phone"
        android:layout_alignLeft = "@ + id /imagecenter"
        android:layout_above = "@ + id /imagecenter"………… />
  …… //省略其他三个 ImageView
```

第一个 ImageView 添加正中间的图片 ESPN,用 layout_centerInParent 设置在容器中央。这个图片将作为其他图片位置的参照物。第二个 ImageView 添加图片 Phone,通过 layout_above+layout_alignLeft,设置图片在参照物图片的左上方。其他三个图片设置请参照表 3 - 3。

表 3 - 3　相对布局中其他三个图片的 XML 设置

图片	XML 设置	设置效果
QQ	android:src="@drawable/qq" android:layout_alignLeft ="@+id/imagecenter" android:layout_below="@+id/imagecenter"/ >	通过 layout_below+layout_alignLeft,设置图片在参照物图片的左下方。

续表

图片	XML 设置	设置效果
Yahoo	android:src="@drawable/yahoo" android:layout_alignTop="@+id/imagecenter" android:layout_toLeftOf="@+id/imagecenter"	通过 layout _ alignTop + layout _ toLeftOf,设置图片和参照物图片顶端对齐且在左边,即为平行的左边。
Youtube	android:src="@drawable/youtube" android:layout_alignTop="@+id/imagecenter" android:layout_toRightOf="@+id/imagecenter"	通过 layout _ alignTop + layout _ toRightOf,设置图片和参照物图片顶端对齐且在右边,即为平行的右边。

图 3-5 使用相对布局技术实现的界面效果

运行程序,效果如图 3-5 所示。

3.2.3 帧布局

帧布局由 FrameLayout 所代表,FrameLayout 直接继承了 ViewGroup 组件。帧布局容器为每个加入其中的组件创建一个空白的区域(称为一帧),每个子组件占据一帧,这些帧都会根据 gravity 属性执行自动对齐。

【微信扫码】
案例 3-6
相关文件

案例 3-6:帧布局的使用

本案例将使用帧布局,实现一个层叠的效果。新建一个 Android 项目,命名为 Layout_Teach_TableLayout。打开 res/layout/activity_main.xml 文件,核心代码如下。

```
<FrameLayout android:id="@+id/framelayout" android:background="#FFF"
        android:foregroundGravity="bottom|right" ……>
    <TextView android:text="深蓝色的背景" android:background="#FF0000FF"
            android:layout_gravity="center" …… />
    <TextView android:text="天蓝色的背景" android:background="#FF0077FF"
            android:layout_gravity="center" …… />
      <TextView android:text="水蓝色的背景"  android:background="#FF00B4FF"
            android:layout_gravity="center" …… />
</FrameLayout>
```

第一个 TextView 为底部的蓝色背景,第二个 TextView 为中间的天蓝色的背景,第三个 TextView 最前面的水蓝色背景,运行程序,效果如图 3-6 所示。

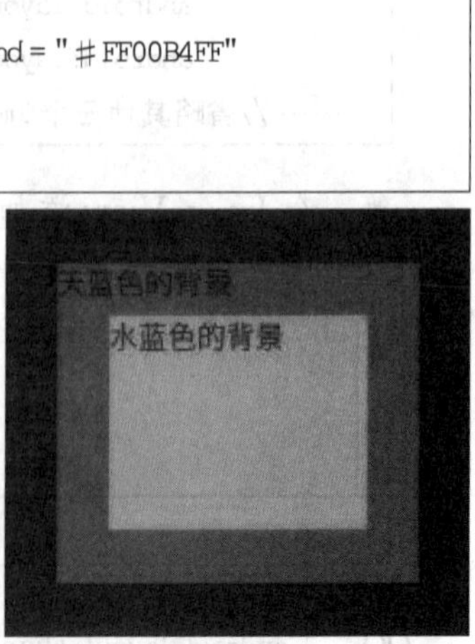

图 3-6 使用帧布局实现的重叠效果

3.2.4 百分比布局

前面的几种布局,都是从 Android 1.0 版本就开始出现了,虽然满足大部分的 UI 设计要求,但当界面上控件比较多,希望通过比例来指定控件大小,而不是设置控件尺寸值的方式来设计界面时,就显得比较困难。例如 LinearLayout

布局，虽然可以通过 layout_weight 属性实现，但只能指定 layout_height 或者 layout_weight 中的一个，不能同时设置两个方向的尺寸，无法实现按比例分配。

为此，Android 引入了一种全新的布局——百分比布局来解决此问题。在这种布局中，可以不再使用 wrap_content、match_parent 等方式来指定控件的大小，而是允许直接指定控件在布局中所占的百分比，这样可以轻松实现任意比例分割布局的效果。

百分比布局通过 FrameLayout 和 RelativeLayout 进行了功能扩展，提供了 PercentFrameLayout 和 PercentRelativeLayout 这两个全新的布局。

案例 3-7：百分比布局的使用

【微信扫码】案例 3-7 相关文件

新建一个项目，命名为 Layout_Teach_PercentFrameLayout。百分比布局定义在了 support 库当中，需要在项目的 build.gradle 中添加百分比布局库的依赖，就能保证百分比布局在 Android 所有系统版本上的兼容性了。

步骤 1：添加依赖。打开 app/build.gradle 文件，在 dependencies 闭包中添加如下内容。

```
implementation 'com.android.support:percent:28.0.0'
```

步骤 2：编写布局文件。修改 activity_main.xml 文件的代码，核心代码如下。

```
<android.support.percent.PercentFrameLayout
    xmlns:App="http://schemas.android.com/apk/res-auto">
  <Button android:id="@+id/button1" android:layout_gravity="left|top"
      App:layout_widthPercent="33%" App:layout_heightPercent="33%"
      android:background="#FF0000" />
  ……//省略其他 5 个按钮
</android.support.percent.PercentFrameLayout>
```

上述代码中，百分比布局并不是内置在系统 SDK 当中，需要把完整的包路径写出来。而且必须定义一个 App 的命名空间，这样才能使用百分比布局的自定义属性。其他 5 个按钮属性设置参照表 3-4。

表 3-4　其他 5 个按钮属性值

XML 属性	Button2	Button3	Button4	Button5	Button6
android:id	button2	button3	button4	button5	button6
android:layout_gravity	right|top	left|center_vertical	center_vertical|right	left|bottom	bottom|right
App:layout_widthPercent	67%	50%	50%	67%	33%
App:layout_heightPercent	33%	33%	33%	33%	33%
android:background	00FF00	0000FF	FFFF00	00FFFF	FF00FF

PercentFrameLayout 继承 FrameLayout 的特性，即所有的控件默认都是摆放在布局的左上角。那么为了让这 6 个按钮不会重叠，这里使用 layout_gravity 来分别将这 6 个按钮放

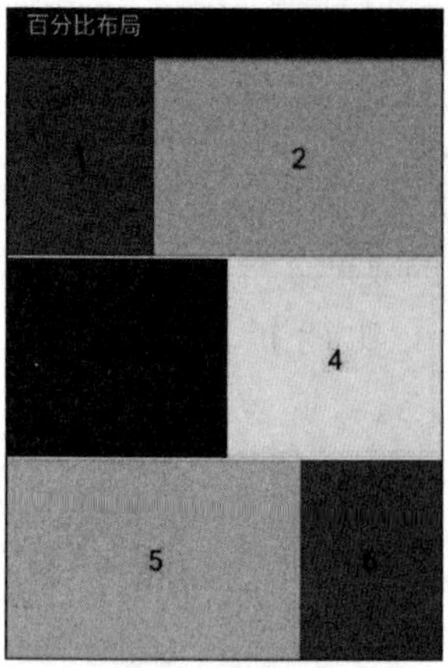

图 3-7　使用百分比布局的效果

置在布局的左上、右上、中左、中右、左下、右下 6 个位置。使用 App:layout_widthPercent 属性将 6 个按钮的宽度指定为布局的 33%、67%、50%、50%、67%、33%。使用 App:layout_heightPercent 属性将各按钮的高度指定为布局的 33%。使用各种颜色来设置按钮背景色，以区分 6 个按钮运行程序，效果如图 3-7 所示。可以看到，6 个按钮都错落有致地分布在界面上，实现了按钮的按比例分配。

3.3　Android 的资源系统

在 Android 应用中，除了 Java 代码之外，还要经常使用字符串、菜单、图像、声音、视频等，上述内容统一称为资源。这些资源，主要是用来定义和显示用户界面的一些静态信息。因此资源是 Android 应用程序中重要的组成部分。

3.3.1　资源类型和存储方式

在 Android 开发中，要求将不同类型的资源进行归类存放。以 XML 文件形式存储的资源可以放在 res 目录中的不同子目录(包括 res/anim、res/color、res/layout、res/menu、res/value 和 res/xml)里。res/raw 目录中可以将任意的资源嵌入 Android 应用程序中，比如音频和视频等。

在 Android Studio 中打开案例 3-7，将左侧的项目资源管理器模式调整到 Project 模式，然后依次点开 app/src/main/res，可以看到项目的资源目录如图 3-8 所示。在 res 目录下，使用不同的子目录来保存不同的应用资源，表 3-5 显示了不同资源在/res 目录下的存储方式。

表 3-5　Android 应用资源的存储

目录结构	资源类型
drawable	该目录下存放各种位图文件，如 png、9.png、jpg、gif 等。
drawable-v24	Android 7.0 以上版本可以将图片资源放在这里。
layout	该目录下存放的是布局文件。
mipmap/anydpi-v26	对于 SDK 版本大于等于 26 的，会采用此文件夹内的自适应图标。
mipmap/hdpi	高分辨率，一般把图片放在这里。
mipmap/mdpi	中等分辨率，很少，除非兼容的手机很旧。
mipmap/xhdpi	超高分辨率，适用手机屏幕材质较好的设备。
mipmap/xxhdpi	超超高分辨率，这个在高端机上有所体现。
values	存放各种简单值的 XML 文件，包括字符串值、整数值、颜色值、数组等。

除此之外，还可以根据需要添加一些资源文件，文件夹的名称和作用如表 3－6 所示。

表 3－6　扩展的 res 资源文件夹

目录结构	资源类型
xml	任意的 XML 文件，可在 Java 中使用 Resources.getXML()方法进行访问。
menu	存放为应用程序定义各种菜单的资源。
anim	存放定义补间动画的 XML 文件。
raw	音频或者视频等原生资源。
assets	任意的原生资源，在应用程序中使用 AssetManager 来访问这些资源。

保存在 res 目录下的资源，Android SDK 会在编译时，通过 R 资源清单类访问这些资源，为这些资源创建索引，并将其自动保存在 R.java 类中，该类包含系统中使用的所有资源文件的标识。资源类包括数组 array、属性 attr、颜色 color、图片 drawable、ID 标识 id、布局 layout、字符串 string 等。R.java 位于 app/build/generated/source/r/debug/<项目包名>/中。文件结构如图 3－9 所示。

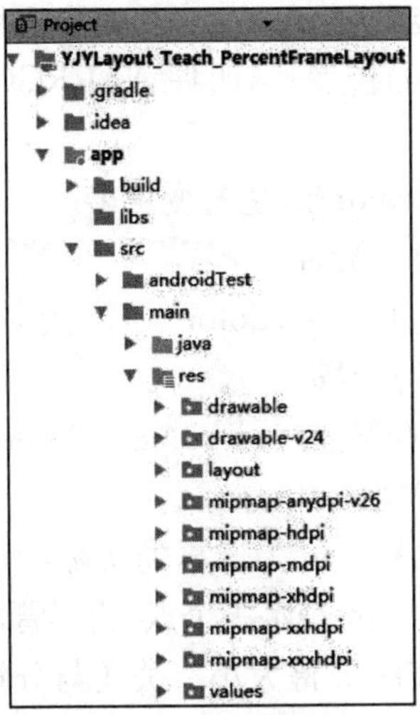

图 3－8　Android 资源的存储目录

```
public final class R {
  public static final class anim {...}
  public static final class attr {...}
  public static final class bool {...}
  public static final class color {...}
  public static final class dimen {...}
  public static final class drawable {...}
  public static final class id {...}
  public static final class integer {...}
  public static final class layout {...}
  public static final class mipmap {...}
  public static final class string {...}
  public static final class style {...}
  public static final class styleable {...}
}
```

图 3－9　R.java 的文件结构

从 R 类中很容易看出，系统为 res 目录中每一个子目录或标签(例如< string >标签)都生成了一个静态的子类，点击查看其中的 class id 和 class string，显示如下。

```
public static final class id { public static final int button1 = 0x7f070022;……}
public static final class string {public static final int App_name = 0x7f0b0027;……}
```

可以看到，用户在项目中定义的 button 和 App_name 都被创建了十六位进制的索引编号。也就是每一个指定 id 属性的组件生成了唯一的 ID，并封装在 id 子类中。这样在 Android 应用程序中可以通过 ID 使用这些组件。

3.3.2 基础资源类型和使用

Android 资源文件使用 XML 来实现,在文件中包含两个部分。第一部分为声明<? xml version="1.0" encoding="utf-8"? >。第二部分为根元素,标签< resourses > </ resourses >。不同类型的资源文件放在根元素中。下面对基础的资源类型和使用进行详细介绍。

(1) 字符串资源 string:定义应用中的一些字符串常量,位置为 res/values/strings.xml。XML 文件的定义为< string name ="字符串的变量名">字符串的值</string >。

如果字符串的值有单引号或者双引号,则需要转换。单引号在外面包上一层双引号,如"abc'def",则值是 abc'def。如果是双引号则需要转义,\"abcd\"则输出"abcd"。

XML 代码中引用资源的格式:@[package:]string/字符串的变量名。

Java 代码中引用资源的格式:R.string. 字符串的变量名。

Java 代码中获取字符串的方法:Resources.getString()。

(2) 颜色资源 color:可以作为 View 组件的背景色、字体颜色等。位置为 res/values/colors.xml。XML 文件的格式为< color name ="颜色的变量名">#颜色值</color >。

Android 中使用 4 个数字来表示颜色,分别是 alpha、红(red)、绿(green)、蓝(blue)。四个颜色值(ARGB)必须以#开头,后面跟着 16 进制的数,有四种表示法。第一种 RGB 形式,如#000 表示黑色;第二种 ARGB 形式,A 表示透明度,A=0 表示完全透明,如#0000 表示透明黑色;第三种 RRGGBB 形式,为第一种形式的扩展;第四种 AARRGGBB 形式,为第二种形式的扩展。

XML 代码中引用颜色资源的格式:@[package:]color/颜色的变量名。

Java 代码中引用颜色资源的格式:R.color.颜色的变量名。

Java 代码中获取颜色的方法:getResources().getColor()。例如:textView.setTextColor(getResource().getColor(R.color. 颜色的变量名))。

(3) 尺寸资源 dimension:用来定义组件的大小,位置为 res/values/dimens.xml,XML 文件格式为< dimen name ="尺寸的变量名">尺寸的大小</dimen >。

尺寸大小的单位有如下形式:px 为屏幕的实际像素。in 为屏幕的实际物理英寸,一英寸等于 2.54 厘米。mm(毫米)为屏幕的实际物理尺寸。pt 表示一个点,为屏幕的物理尺寸,大小是 1/72 英寸。dp 是一个与密度无关的像素,与实际屏幕大小有关还与分辨率有关,这个单位最为常用。sp 是与比例无关的像素,与 dp 类似,但是除了适应屏幕密度外还适合用户的字体,安卓官方建议在设置 textSize 的时候使用该单位。

XML 代码中引用尺寸资源的格式:@[package:]demin/尺寸的变量名。

Java 代码中引用尺寸资源的格式:R.string.尺寸的变量名。

Java 代码中获取尺寸的方法:Resources. getDimension(),例如:float dimension = getResources().getDimension(R.dimen.尺寸名称)。

(4) 数组资源 array:位置为/res/values/arrays.xml,XML 文件的格式如下。

```
子元素<数组类型 - array name = "数组的变量名">
子项< item >数组元素的值< /item >
```

其中数组类型可包含如下三种子元素:< array…/ >定义普通类型的数组;< string-array…/>定义字符串数组；<integer-array…/>定义整数数组。XML 不能引用数组类型的资源。

Java 代码中获取数组类型的方法：getResource().getStringArray(R.array.数组的变量名)。

(5) 图片资源 drawable：保存的是图片资源，保存的格式常用的有.bmp，.png，.gif，.jpg 等。文件名必须由英文或数字组成，drawable 又可分为位图文件(bitmap file)、颜色(color drawable)、九图图片(nine-patch image)等，位置为 res/drawable。

XML 代码中引用 drawable 资源的格式：@[package:]drawable/图片的变量名。

Java 代码中引用 drawable 资源的格式：R.drawable.图片的变量名。

Java 代码中获取 drawable 资源的方法：getResources().getDrawable(R.drawable.图片变量名)。

获取之后返回的都是 Drawable 对象。图像文件返回 BitmapDrawable 对象。9.png 图像返回 NinePatchDrawable 对象。

下面通过一个案例，介绍资源文件的使用。

【微信扫码】案例 3－8 相关文件

案例 3－8：基础资源文件的综合运用

在这个案例中，将综合使用字符串、尺寸和颜色等资源设计一个多彩面板。新建一个 Android 项目，命名为 Res_Teach_ResStringColDimXML。

步骤 1：编写尺寸资源文件。打开 app/res/values/dimens.xml 文件，添加核心代码如下。

```
<dimen name = "dimen1"> 24dp </dimen>……
```

以此类推，分别定义 36dp，48dp，60dp，72dp，84dp，96dp 几个尺寸。

步骤 2：编写颜色资源文件。打开 app/res/values/colors.xml 文件，添加颜色资源代码如下。

```
<color name = "color1">#FF0000 </color>……
```

以此类推，分别定义#FFA500，#FFFF00，#00FF00，#00FFFF，#0000FF，#FF00FF。

步骤 3：编写字符串资源文件。打开 app/res/values/strings.xml 文件，编写代码如下。

```
<string name = "strcolor1">赤</string>
```

以此类推，分别定义橙，黄，绿，青，蓝，紫其他几个颜色资源。

步骤 4：编写布局文件。打开 app/res/layout/activity_main.xml 文件，编写代码如下。

```
<LinearLayout android:layout_height = "match_parent"
      android:layout_width = "match_parent"   android:orientation = "vertical">
    <TextView android:id = "@ + id /str1"   android:text = "@string /strcolor1"
        android:background = "@color /color1"   android:layout_width = "match_parent"
        android:layout_height = "@dimen /dimen1"   />
        …… //省略其他 6 个 TextView
</LinearLayout>
```

布局文件在垂直方向上放置 7 个 TextView。每个文本框的字体颜色、背景色和文本框的宽度分别使用资源文件中定义的属性值。

步骤 5:编写实现代码。打开 app/java/包名/MainActivity.java,编写代码如下。

```
public class JAVAMainActivity extends Activity {
  private TextView[ ] textview = new TextView[7]; //声明使用 7 个的 TextView
  private int[ ] textstr = new int[ ]{R. string. strcolor1, ……};
  private int[ ] color = new int[ ]{R. color. color1 ……};
  private int[ ] dimen = new int[ ]{R. dimen. dimen1, ……};
   @Override
   protected void onCreate(Bundle savedInstanceState) {
   super. onCreate(savedInstanceState);setContentView(R. layout. java_activity_main);
   LinearLayout layout = (LinearLayout)findViewById(R. id. linearlayoutjava);
   for (int i = 0;i<textview. length;i++){  //用循环的方式来生成组件
       textview[i] = new TextView(this);  //创建一个 TextView 对象
       textview[i]. setGravity(Gravity. CENTER);  //设置文本居中
       textview[i]. setText(textstr[i]);  //设置文本框显示的文字
       //设置文本框的背景色和宽度尺寸
       textview[i]. setBackgroundColor((int)(super. getResources(). getColor(color[i])));
       textview[i]. setHeight((int)(super. getResources(). getDimension(dimen[i])));
       layout. addView(textview[i]); //将文本框组件添加到线性布局中 } } }
```

运行程序,可以看到 7 个文本框分别使用了不同的字、背景色和尺寸,显示出彩色面板的效果,如图 3 - 10 所示。

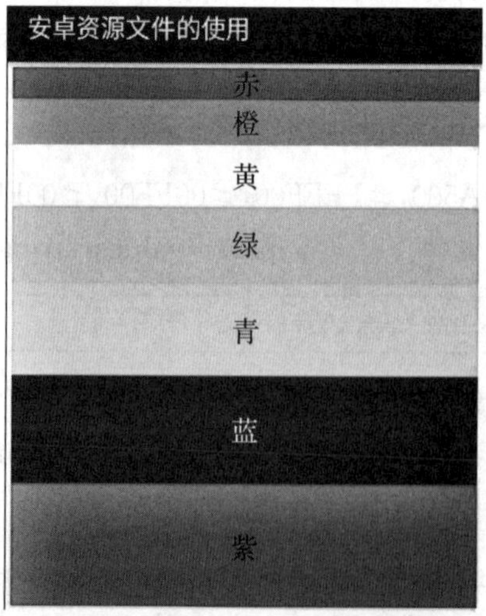

图 3 - 10　基础资源文件案例运行效果

3.3.3　高级资源类型和使用

(1) 主题资源 style:相当于网页设计中的样式表。它也是一种资源类型,用来统一 UI 界面的风格。位置为 res/values/styles.xml,XML 文件格式如下。

```
< style name = "主题变量名">
< style name = "新的主题变量名"  parent = "继承的父类主题变量名" >
< item name = "属性名">属性值< /item >
```

XML 代码中引用资源的格式:@ R.style.主题变量名。

Java 代码中引用资源的格式:在每个组件内部的属性设置里 style ="@/style/主题变量名。

在 AndroidMainfest.xml 文件里的 application 标签里使用:Android: theme ="@style/主题名称"。

Android 系统自带了相当丰富的主题资源,在界面设计时也可以引用 Android 自带的主题资源,语法格式:? [package:]type/主题变量名。

引用 style 属性的语法与引用资源的语法类似,区别只是将@标记换成了"?"。android:textColor=="? Android:textDisabledColor"。

(2) 布局资源 layout:布局资源是 Android 中最常使用的一种资源,从我们学习第一个 Android 应用开始,就已经开始接触 Android 的布局资源了。位置在 res/layout/文件夹中,XML 文件格式如下。

```
<布局类 xmlns:android = http: //schemas.android.com /apk /res /android (属性)>
<视图组件或其他嵌套布局类>< /…>< /布局类>
```

XML 代码中引用布局资源的格式:@[package:]layout/布局文件变量名。需要注意在布局资源中不可以嵌套使用,就是不能自己引用自己。

Java 代码中引用布局资源的格式:R.layout.布局文件变量名。

Java 代码中获取字符串的方法:Activity.setContentView(R.layout.布局文件变量名)。

(3) raw 和 assets 资源:两者都可以保存任何类型的文件,可以被封装到 apk 里,但是不会被编译。在布局中不可以使用,在 Java 代码中使用方法如下。

```
InputStream read =  getResources().openRawResource(R.raw.文件名)
OutputStream write =  getResources().openRawResource(R.raw.文件名)
```

或者

```
InputStream read = getAssets().open("文件的名称")
OutputStream write = getAssets().open("文件的名称")
```

(4) 菜单资源 menu:在 res/menu 目录下,以< menu >为根标签,XML 结构如下。

```
< menu 命名空间>
    < item 设置的属性 />
    < group id >< item 设置的属性 />< menu >< item 设置的属性 />< /menu >< /group >
< /menu >
```

menu 标签没有任何属性,除了命名空间,其他不要。命名空间可以是安卓默认的也可以由自己定义。菜单的 item 属性和 group 的属性如表 3-7 和表 3-8 所示。

表 3-7 菜单的 item 属性

属性名	属性描述
id	菜单项的 ID
menuCategory	菜单项的种类，如设置成 system 表示系统菜单
orderInCategory	同种类菜单排列顺序
title	菜单项的显示文本
titleCondensed	菜单项的短标题，如果菜单项文本太长，会显示该值
icon	菜单项图片的 ID
alphabeticShortCut	菜单项的字母快捷键
numericShortCut	菜单项数字快捷键
checkable	菜单项是否带复选框
checked	如果菜单项带复选框，表示该复选框是否被默认选中
visible	菜单项是否可见
enabled	菜单项是否可用

表 3-8 菜单的 group 属性

属性名	属性描述
id	菜单组的 ID
menuCategory	与 item 相同，只是作用域在菜单组
orderInCategory	与 item 相同，只是作用域在菜单组
visible	菜单组里的所有菜单项是否可见
enable	菜单组里所有菜单是否可用
CheckableBehavior	设置该菜单组上显示的选择组件

菜单文件在布局文件中不可以使用。Java 代码中获得菜单的方法如下。在 onCreateOptionsMenu(Menu menu)或 onCreteContextMenu()的回调方法里装载。

```
MenuInflater menuinflater = getMenuInflater()
menuinflater.inflate(R.menu.菜单资源文件的名称)
```

如果是 onCreteContextMenu 里，要在 onCreate()方法里将上下文菜单注册到某个组件上如 registerForContextMenu(某个组件的名称)。

(5) 动画资源 anim：在代码中如果输入 android.R.anim.会弹出图 3-11 所示 Android 系统自带的 anim 动画效果。在 Java 代码中使用 loadAnimation 加载动画效果，格式如下。

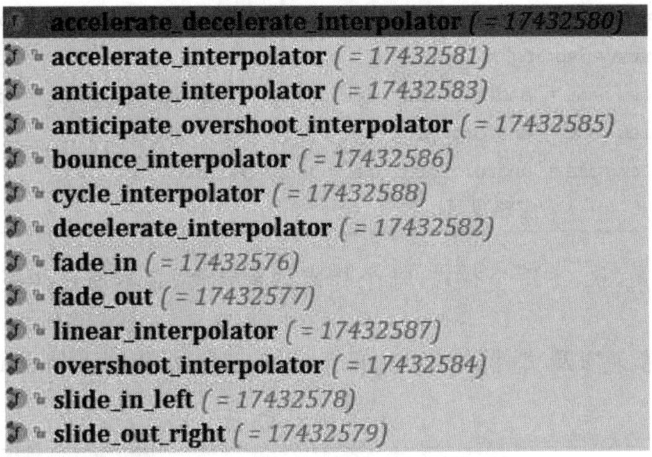

图 3－11　Android 系统自带的 anim 效果

```
public static Animation loadAnimation (Context context, int id)
```

其中参数 context 为程序的上下文，参数 id 为动画资源名。例如 myAnimation = AnimationUtils.loadAnimation(this,R.anim.动画资源名)。

如果在 imageView 中使用动画资源，格式如下。

```
ImageView imageView = (ImageView)findViewById(R.id.图像组件的 id)
imageView.setBackgroundResources(R.anim.动画资源名)
```

下面我们将使用高级资源实现一个简单的切换案例。

案例 3－9：使用主题和样式的案例

在本案例中，我们将使用不同的布局、样式和主题等资源文件，实现一个能显示不同季节风格且可以自动切换图片程序。新建一个 Android 项目命名为 Res_Teach_StyleChange。

【微信扫码】
案例 3－9
相关文件

步骤 1：编写尺寸资源文件。打开 app/res/values/dimens.xml 文件，编写代码如下。

```
<dimen name = "bntsummer"> 56dp </dimen>
…… //以此类推，添加 40dp、48dp、24dp 几个尺寸
```

步骤 2：编写字符串资源文件。打开 app/res/values/strings.xml 文件，编写代码如下。

```
<string name = "springtext">春天的脚步在哪里?在心里，在脑海里</string>
<string name = "summertext">炎炎夏日，骄阳似火</string>
…… //添加若干个
```

步骤 3：编写样式资源文件。打开 app/res/values/styles.xml 文件，编写代码如下。

```
< style name = "AppTheme"parent = "android:Theme.Material.Light.DarkActionBar">< /style >
    < style name = "spring">
        < item name = "android:windowBackground">@drawable /spring < /item >
        < item name = "android:textColor">#FF0000 < /item >
        < item name = "android:textSize"> 36sp < /item >
    < /style >  …… //省略夏、秋、冬
```

分别设置春、夏、秋、冬四个样式,每个样式的背景色、文本颜色和字体大小都不一样即可。

步骤 4:编写春天布局文件。打开 app/res/layout/activity_main.xml 文件,编写代码如下。

```
< LinearLayout android:id = "@ + id /linearlayout" android:orientation = "vertical" ……>
    < TextView android:id = "@ + id /text1"
            android:text = "@string /springtext" style = "@style /spring" …… />
    < Button android:id = "@ + id /bntchange" android:layout_gravity = "bottom|center"
                android: text = " @ string / bntspring " android: textSize = " @ dimen /
bntspring" …… />
< /LinearLayout >
```

步骤 5:编写其他的布局文件。在文件夹 app/res/layout 中,分别新建布局文件 summer_activity_main.xml、autumn_activity_main.xml、winter_activity_main.xml 作为其他三季的布局文件,实现代码参考春天的布局文件。

步骤 6:编写实现代码。打开 app/java/包名/MainActivity.java 文件,编写代码如下。

```
public class MainActivity extends Activity {
   private int[] theme = new int[]{R.style.spring, R.style.summer, R.style.autumn, R.style.
winter}; //声明需要的资源文件——样式、布局文件、图片、按钮
   private int[] activitymain = new int[]{R.layout.activity_main, R.layout.summer_activity_
main, R.layout.autumn_activity_main, R.layout.winter_activity_main};
   private int[] resimage = new int[]{R.drawable.spring, R.drawable.summer, R.drawable.
autumn, R.drawable.winter};
   private Button bntchange = null;private View view = null;private int i = 0;
   @Override
   protected void onCreate(Bundle savedInstanceState) {
     super.setTheme(R.style.spring); //加载布局文件之前就需要调用的 style,这点非常重要
     super.onCreate(savedInstanceState);    setContentView(R.layout.activity_main);
     //获得按钮的对象,并设置按钮的点击事件
     this.bntchange = (Button) super.findViewById(R.id.bntchange);
     this.bntchange.setOnClickListener(new OnBntChClickListenerImpl());}
```

加载布局文件,加载的是春天的布局文件。写到这里,运行程序出来的第一个界面,是春天的布局文件,但是点击按钮是没有反应的,需设置按钮的事件。

步骤 7:编写按钮点击事件代码。每单击一次按钮将切换到相关界面信息,代码如下。

```
private class OnBntChClickListenerImpl implements View.OnClickListener{
    @Override
    public void onClick(View v) {
        i = i + 1; //获得当前的界面是第几个界面
        if (i = = 4) i = 0; //设置循环模式,防止在最后的界面切换到下一个界面时出错
        MainActivity.this.setTheme(theme[i]); setContentView(activitymain[i]);
        view = findViewById(R.id.linearlayout); view.setBackgroundResource(resimage[i]);
        bntchange = (Button) findViewById(R.id.bntchange);
        //给新的界面生成按钮点击事件,否则无法在使用新的按钮,这点非常重要
        bntchange.setOnClickListener(new OnBntChClickListenerImpl()); } } }
```

运行程序效果如图 3-12 所示,四季的背景、布局、文字颜色等都不一样,可以循环切换。

图 3-12　使用主题和样式的效果——春、夏、秋、冬

小　结

本章主要讲解了 Android 中的界面设计、布局和资源类型等,这些是 Android 开发的基础,所有的 Android 程序都需要用到,因此要求读者必须熟练掌握,为后面的学习做好铺垫。

课后练习

【微信扫码】
第 3 章课后练习

第 4 章

基本 UI 组件

界面是由控件组合而成的，想要构建功能强大的界面，需要学习相关控件的知识，了解这些控件的属性和方法并加以灵活运用。

Android Studio 和 Eclipse 中都提供了控件的可视化编辑器，允许使用拖放控件的方式来添加控件并直接修改控件的属性。不过一般不推荐使用这种方式，因为其可视化编辑工具并不利于真正了解界面背后的实现原理。通过这种方式制作出的界面通常不具备良好的屏幕适配性，当需要编写较为复杂的界面时，将很难胜任。因此推荐使用最基本的方式去实现，即编写 XML 代码。

在 Java 中，所有的类都是 Object 的子类，Android 属于 Java，因此 Android 的类全部继承于 Object，在 Android 中，全部控件都是 View 和它的子类。下面就介绍一些常用的控件的使用。

4.1 TextView 类组件

TextView 类是 View 的子类，是 Android 中最简单也是使用最广泛的一个控件，前面一章里，我们已经和它打过交道了，本章继续深入学习，掌握它更多的用法。

TextView 直接子类有 Button，CheckedTextView，Chronometer，DigitalClock，EditText，TextClock 等控件。间接子类有 AutoCompleteTextView，CheckBox，CompoundButton，ExtractEditText，MultiAutoCompleteTextView，RadioButton，Switch，ToggleButton 等控件。

可见，TextView 类是很多控件的父类，它的很多属性和方法可以被这些子类继承。

4.1.1 TextView 和 EidtView

TextView(文本框)不能接收用户的单击等事件消息，只能用来显示静态文本。表 4-1 列出了 TextView 控件的一些常用属性。

表 4-1 TextView 的常用属性

属性	描　述
android:text	设置要显示的文本信息。
android:layout_width	代表组件宽度。
android:layout_height	代表组件高度。
android:textSize	代表字体大小。
android:textColor	代表字体颜色。
android:background	代表背景。

续表

属性	描　　述
android:padding	代表边距。
android:layout_margin	代表边界。
android:antolink	根据特定的掩码模式来判断，当符合指定模式时，将自动为其加上链接。默认情况下 none 是不进行任何内容匹配，其他包括：phone——系统的拨号程序；email——电子邮件发送程序；map——地图浏览程序；web——浏览器浏览 URL 网页；all——调用任何符合现有的应用程序来访问对应的数据信息。

EditView（编辑框）主要用来获取用户的输入信息。常用属性如表 4－2 所示。

表 4－2　EditView 常用属性

属性	描　　述
android:maxLength	最大输入字符数，比如 maxLength=4 表示最多能输入 4 个字符。
android:hint	显示在输入文本框内的提示信息。
android:numeric	输入数字的类型，其中 integer 设置只能输入整数，decimal 设置允许输入小数。
android:singleLine	是否为单行输入，true 为单行输入，文字不会自动换行。
android:typeface	输入文本的字体，允许值有 normal，sans，serif，monospace。
android:inputType	文本的类型，让输入法选择合适的软键盘。常用的允许值有 phone，date，datetime，time，number 等。

下面通过一个案例介绍 TextView 和 EditView 的使用。

案例 4－1：TextView 和 EditView 的使用

新建一个 Android 项目，命名为 UI_Teach_TextEditView。在文件夹 app/res/layout 中，修改布局文件 activity_main.xml，在界面中放置几个编辑框，分别用来输入用户名、密码和年龄。用户名和密码需要字母和数字，年龄只能是数字，通过设置不同的 InputType 来进行区分。代码如下。

【微信扫码】
案例 4－1
相关文件

```
<LinearLayout android:id="@+id/linearlayout1" android:orientation="vertical"……>
    <TextView android:text="用户名(最大 6 位):"……  />
    <EditText android:id="@+id/editusername" android:maxLength="6"……  />
    <TextView android:id="@+id/textpassword" android:text="你的密码:" ……/>
    <EditText android:id="@+id/editpassword" android:inputType="textPassword" ……/>
    <TextView android:id="@+id/textpasswordagain" android:text="重复密码:" ……/>
    <EditText android:id="@+id/editpasswordagain" android:inputType="textPassword"
        android:hint="@string/stringhint"   />
    <TextView android:id="@+id/textemail" android:text="邮箱地址(username@sina.
com):" android:autoLink="email" …… />
```

```
    <EditText android:id="@+id/editemail" android:inputType="textWebEmailAddress"…… />
    <TextView android:id="@+id/textage" android:text="年龄:" …… />
    <EditText android:id="@+id/editeage" android:inputType="numberSigned"…… />
</LinearLayout>
```

编辑完成之后，无须修改 MainActivity 的代码直接运行程序，分别点击不同的 EditView 中，如图 4-1 和 4-2 所示，可以看到在不同的编辑框中，弹出的键盘类型不同，效果实现不同。

图 4-1　输入用户名的界面

图 4-2　输入年龄的界面

4.1.2　Button

Button(按钮)的作用是当用户单击时，响应用户输入执行特定的操作。在使用中可以通过两种方式来完成事件响应。第一种方式，先注册事件监听器并通过其回调方法来实现。第二种方式，在 XML 布局文件中，用 android:onClick 属性为按钮指定方法名，以代替使用 OnClickListener 事件监听模式。下面介绍 Button 的使用。

【微信扫码】
案例 4-2
相关文件

案例 4-2:Button 的使用

新建一个 Android 项目，命名为 UI_Teach_Button。

步骤 1:在布局文件 activity_main.xml 中，在界面中放置三个按钮，核心代码如下。

```
<Button android:id="@+id/login" android:text="登录" …… />
<Button android:id="@+id/register" android:text="注册" …… />
<Button android:id="@+id/reset" android:onClick="myClick" android:text="重置" …… />
```

第三个重置按钮使用 onClick 属性来设置相应单击的相应操作。界面如图 4-3 所示。

图 4-3　Button 设计界面

步骤 2:打开 app/java/包名/MainActivity.java 文件，修改代码如下。

```
public class MainActivity extends Activity {
    @Override
    protected void onCreate(Bundle savedInstanceState) {
      …………
      Button login = (Button)findViewById(R.id.login); //找到登录按钮控件
      Button bntregister = (Button)findViewById(R.id.register); //找到注册按钮控件
      login.setOnClickListener(new View.OnClickListener() {
      @Override
      public void onClick(View v) { //注册事件监听器,实现其中的主要方法 onClick
         Toast toast = Toast.makeText(MainActivity.this, "您单击了"登录"按钮", Toast.
LENGTH_SHORT);toast.show(); } });    }
```

图 4 - 4　单击 Button 的显示效果

当单击登录按钮,显示信息。运行效果如图 4 - 4 所示。

步骤 3:实现注册按钮的功能。首先在 onCreate()中添加下面代码。

```
bntregister.setOnClickListener(new OnClickListenerImpl());
```

代码表示将事件监听器独立出来,作为一个单独的类 OnClickListenerImpl 来处理。然后在 MainActivity 中,新增实现 OnClickListenerImpl 的类,代码如下。

```
private class OnClickListenerImpl implements View.OnClickListener {
    @Override
    public void onClick(View v) {Toast toast = Toast.makeText(MainActivity.this, "您单击
了"注册"按钮", Toast.LENGTH_SHORT); toast.show();}}
```

implements 表示继承于 View.OnClickListener,要实现其中的 onClick 方法。

步骤 4:实现重置按钮的功能。因为在布局文件中 android:onClick="myClick",具体实现是在 Activity 中新增一个 public 类型的以 View 组件为参数的方法,代码如下。

```
public void myClick (View view) { Toast toast = Toast.makeText(MainActivity.this, "您单击了
"重置"按钮", Toast.LENGTH_SHORT); toast.show(); }}
```

上面介绍了 Button 的三种事件监听器的实现方法。总结一下,方法一比较直接和简单,但是缺乏灵活性。当几个按钮都需要设置为同样的效果时,例如计算器的数字按钮,使用方法二比较方便。当按钮数量比较多,而每个按钮单击事件监听功能都不一样,使用方法三会很方便。

4.1.3　RadioButton 和 CheckBox

在现实生活中我们需要进行选择操作,这种选择分为单选和复选两种。

单选操作使用 RadioButton(单选按钮),它有选中和未选中两种状态,当处于未选中状

态时，用户可以通过按下或点击来选中它，一旦选中了就不能取消选中。RadioGroup 是单选钮的容器，RadioButton 必须和 RadioGroup 一起使用，一个 RadioGroup 中的所有 RadioButton 只能有一个被选中，选中其中一个的同时将取消其他 RadioButton 的选中状态。RadioGroup 的属性如表 4 - 3 所示。

表 4 - 3　RadioGroup 的属性

方　法	描　述
public void check (int id)	设置选中的单选按钮的编号
public void clearCheck ()	清空选中状态
public int getCheckedRadioButtonId()	取得选中的单选按钮的 ID
public void setOnCheckedChangeListener(RadioGroup.OnCheckedChangeListener listener)	设置单选按钮选中的操作事件

复选操作使用 CheckBox(复选框)，有选中和未选中两种状态，但可以选择多个同时选中。例如个人兴趣爱好一定是多种的，当需要用户进行选择时候，使用 CheckBox 一次性选择多个即可完成功能。下面通过一个案例来学习两种按钮的使用。

案例 4 - 3：RadioGroup 和 CheckBox 的综合运用

本案例将使用单选按钮和多选按钮，实现一个用户信息注册的界面。新建一个 Android 项目，命名为 UI_Teach_CheckBoxRadioGroup。

步骤 1：修改布局文件 app/res/layout/activity_main.xml，如图 4 - 5 所示，代码如下。

【微信扫码】
案例 4 - 3
相关文件

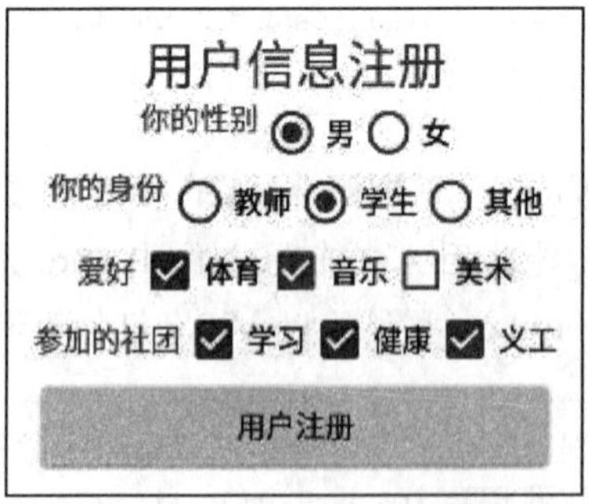

图 4 - 5　用户信息注册界面

```
< TableLayout android:id = "@ + id /tablelayout" ……>
  < TableRow android:id = "@ + id /tablerow1" ……>
        < TextView android:id = "@ + id /textview1" android:text = "用户信息注册" …… />
  < /TableRow >
  < TableRow android:id = "@ + id /tablerow2" …… >
```

```
        < LinearLayout android:gravity = "center_horizontal" ······>
          < TextView android: id = "@ + id /textview2"    android: text = "你的性别" ······ />
          < RadioGroup android: id = "@ + id /radioGroupSex"······>
              < RadioButton android: id = "@ + id /radioman" android: text = "男" ······ />
              < RadioButton android: id = "@ + id /radiowoman" android: text = "女"······ />
          < /RadioGroup >
      < /LinearLayout >
   < /TableRow >
   ······ //省略,内容为你的身份,包括教师、学生、其他等单选项
  < TableRow android: id = "@ + id /tablerow4" ······ >
     < LinearLayout android:gravity = "center_horizontal"······>
       < TextView android: id = "@ + id /textview4" android: text = "爱好" ······ />
         < CheckBox android: text = "体育"    android: id = "@ + id /like1" ······ />
         ······ //省略、内容为音乐、美术等复选选项
     < /LinearLayout >
   < /TableRow >
   ······ //省略,内容为参加的社团复选项:包括学习、健康等复选项
   < TableRow android: id = "@ + id /tablerow6" ······>
         < Button android: id = "@ + id /bntsubmit"    android: text = "用户注册" ······  />
   < /TableRow >
< /TableLayout >
```

步骤 2:打开 app/java/包名/MainActivity.java,修改代码如下。

```
public class MainActivity extends Activity {
    ······ //声明程序需要的控件和数组等
    @Override
    protected void onCreate(Bundle savedInstanceState) {
        ······ //获得布局中的布局和控件
        bntsubmit = (Button) findViewById(R. id. bntsubmit); //添加按钮的监听事
        bntsubmit. setOnClickListener(new OnClickListenerImpl()); }
```

步骤 3:实现按钮的单击事件,在 MainActivity 中新增类 OnClickListenerImpl,代码如下。

```
public class OnClickListenerImpl implements View. OnClickListener { //按钮监听事件
   @Override
   public void onClick(View v) {
     int k = 0; str[k] = "您的选择是"; String strshow = "";
     for (int i = 0;i < rgsex. getChildCount();i ++ ){ //两个单选按钮组的选择
       RadioButton rbsex = (RadioButton) rgsex. getChildAt(i);
       if (rbsex. isChecked()){
              str[ ++ k] = " 您的性别是:" + rbsex. getText(). toString();break;}}
```

```
        for (int i = 0;i < rgidentity.getChildCount();i ++ ){
          RadioButton rbidentity = (RadioButton)rgidentity.getChildAt(i);
          if (rbidentity.isChecked()){
            str[ ++ k] = " 您的身份是:" + rbidentity.getText().toString();break;}}
      str[ ++ k] = "您的爱好是:"; //获得复选的选择
      if (checkbox1.isChecked()) str[k] + = checkbox1.getText().toString() + " ";
      …… //依照 checkbox1,依次获得 checkbox2 - 6 的选择内容
      for (k = 0;k < str.length;k ++ ){strshow + = str[k] + "\r\n";} //将字符串数组转换
为 String
      Toast.makeText(MainActivity.this, strshow, Toast.LENGTH_LONG).show(); } } }
```

编写完成之后,运行程序,效果如图 4 - 6 所示。

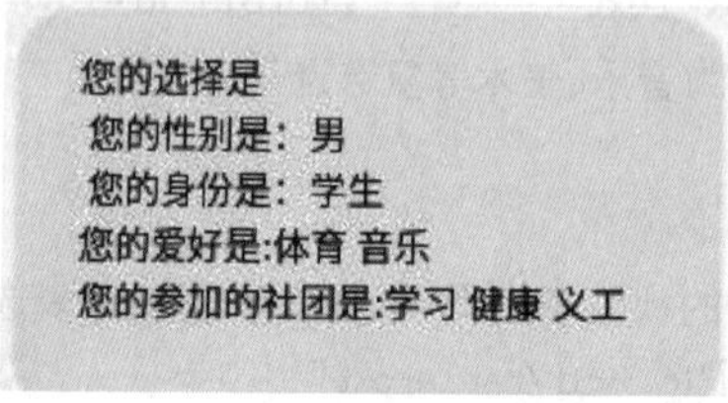

图 4 - 6　用户信息注册显示效果

4.2　ProgressBar 类组件

ProgressBar 类组件用于在界面上显示各种进度的信息,按照显示的方式不同,分为进度条、星级评分条和拖动条三种,类的构成如图 4 - 7 所示。

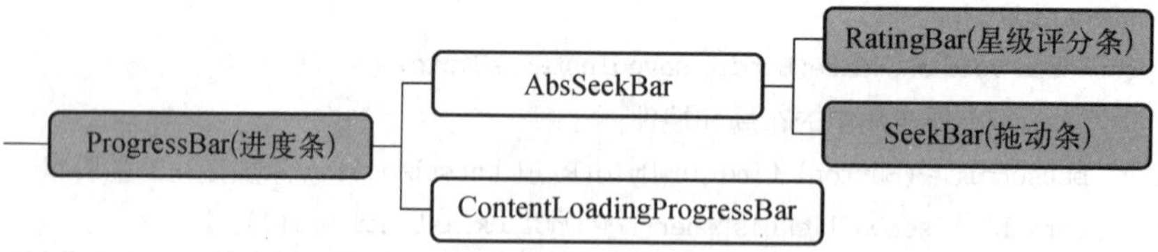

图 4 - 7　ProgressBar 类的构成

4.2.1　ProgressBar

ProgressBar(进度条)分为两类,一类是有明确进度信息的,将通过一个带刻度的水平条来表示,另一类是没有明确进度信息的,可以通过一个动画图标来表示,它通过调用 requestWindowFeature (Window.FEATURE_PROGRESS)打开窗口的进度条特性。

4.2.2　SeekBar

SeekBar(拖动条),作为一种直观的输入控件,可作为音乐播放器等的播放进度指示,音量调整工具等。SeekBar 是 ProgressBar 的扩展,在其基础上增加了一个可滑动的滑片。

用户可以触摸滑片向左或向右拖动,或者可以使用方向键设置当前的进度等级。

调用 setMax、setProgress 等设置 SeekBar 组件的属性,实现 OnSeekBarChangeListener 接口,利用 onProgressChanged、onStartTrackingTouch 和 onStopTrackingTouch 等方法监听 SeekBar 组件与用户的交互,也可以附加一个 SeekBar.OnSeekBarChangeListener 以获得用户操作的通知。主要的方法如下。

(1) public abstract void onProgressChanged (SeekBar seekBar, int progress, Boolean fromUser):进度已经被修改。

参数 seekBar:当前被修改进度的 SeekBar。

参数 progress:当前的进度值。此值的取值范围为 0 到 max 之间,max 默认为 100,用户可以通过 setMax(int) 设置。

参数 fromUser:区分触发改变的类型,用户触发返回 true,编程触发返回 false。

(2) public abstract void onStartTrackingTouch (SeekBar seekBar):用户已经开始一个触摸拖动手势。

(3) public abstract void onStopTrackingTouch (SeekBar seekBar):用户触摸手势已经结束。

4.2.3 RatingBar

RatingBar(星级评分条)是基于 SeekBar 和 ProgressBar 的扩展,用星型来显示等级评定。使用 RatingBar 时,用户可以触摸、拖动或使用方向键来设置评分,它有小样式和大样式两种显示方式。

使用星级评分条,需要创建 OnRatingBarChangeListener 并将其注册到 Ratingbar 的事件监听器中,在 OnRatingBarChangeListener 中需要实现 onRatingChanged 方法。

方法 onRatingChanged (RatingBar ratingBar, float rating, Boolean fromUser)在评分等级已被修改时调用。用户拖动评分条时不会被调用,仅仅当用户使用触摸方式修改评分等级,修改结束之后调用。参数 ratingBar 为评分修改的星级。Rating 为当前分数,取值范围为 0 到星型的数量。fromUser 为评分改变是由用户触摸手势、方向键或轨迹球移动触发的。

下面通过案例介绍星级评分条的使用,进度条和拖动条由读者自行完成。

案例 4-4:RatingBar 的使用

在 Android 中新建一个项目,命名为 UI_Teach_RatingBar,实现星级评分条的使用,并区分是由何种方式触发评分操作的。

【微信扫码】
案例 4-4
相关文件

步骤 1:在文件夹 app/res/layout 中,修改布局文件 activity_main.xml,代码如下。

```
<RatingBar android:id="@+id/ratingBar1" android:numStars="5"
        android:rating="3.5" android:isIndicator="false" …… />
<Button android:text="提交" android:id="@+id/button1"…… />
<Button android:text="改变" android:id="@+id/button2" …… />
```

界面上布置星级评分条和两个按钮。效果如图 4-8 所示。

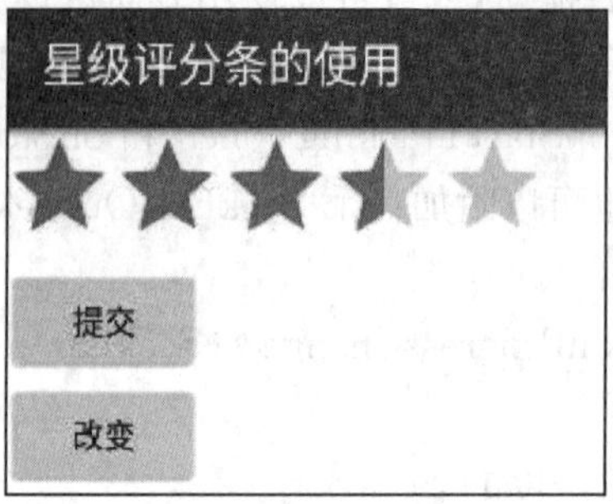

图 4-8 星级评分条的设计界面

步骤 2:打开 app/java/包名/MainActivity.java 文件,直接点击星级评分条,实现修改评分功能,核心代码如下。

```
public class MainActivity extends AppCompatActivity {
    private Toast toast;private RatingBar ratingbar;    //星级评分条
    @Override
    protected void onCreate(Bundle savedInstanceState) {
      super.onCreate(savedInstanceState); setContentView(R.layout.activity_main);
      ratingbar = findViewById(R.id.ratingBar1);
  ratingbar.setOnRatingBarChangeListener(new RatingBar.OnRatingBarChangeListener(){
    @Override
    public void onRatingChanged(RatingBar ratingBar, float rating, boolean fromUser) {

      tring toaststr = "自行改变" + "\r\n";
      if (fromUser = = true){toaststr = "用户触摸改变" + "\r\n"; } //判断如何改变评分条
状态
     toaststr + = "你改变到" + rating + "颗星";
      toast = Toast.makeText(MainActivity.this, toaststr, Toast.LENGTH_SHORT);
     toast.setGravity(Gravity.CENTER, 0, - 200); toast.show(); } });
```

修改效果如图 4-9 所示。

图 4-9 星级评分条的触摸改变效果

步骤 3:实现提交按钮的功能,将当前星级评分条的数据显示出来。

```
Button button = (Button)findViewById(R.id.button1); //获取提交按钮
button.setOnClickListener(new View.OnClickListener() {
  @Override
   public void onClick(View v) {
     int result = ratingbar.getProgress();float rating = ratingbar.getRating();
     float step = ratingbar.getStepSize();
     toast = Toast.makeText(MainActivity.this, "你得到了" + rating + "颗星", Toast.LENGTH
_SHORT);toast.setGravity(Gravity.CENTER, 0, - 200); toast.show();}  });   }}
```

步骤 4:实现改变按钮的功能,以每次 1/2 颗星累加。

```
Button bntchange = (Button)findViewById(R.id.button2);
bntchange.setOnClickListener(new View.OnClickListener() {
    @Override
    public void onClick(View v) { //设置当前进度、每次最少改变星级、获取当前等级
      int result = ratingbar.getProgress();float step = ratingbar.getStepSize();
      float rating = ratingbar.getRating();result + = step;rating + = step;
        ratingbar.setProgress(result);ratingbar.setRating(rating);} });
```

提交按钮显示效果图 4 - 10 所示,改变按钮效果如图 4 - 11 所示。

图 4 - 10　星级评分条的提交效果

图 4 - 11　星级评分条的按钮点击改变效果

4.3　ViewAnimator 类组件

ViewAnimator 是一个 FrameLayout 的基类,可以在视图之间进行切换,实现动画效果,它派生出切换视图和翻转视图等几个控件,关系如图 4 - 12 所示。

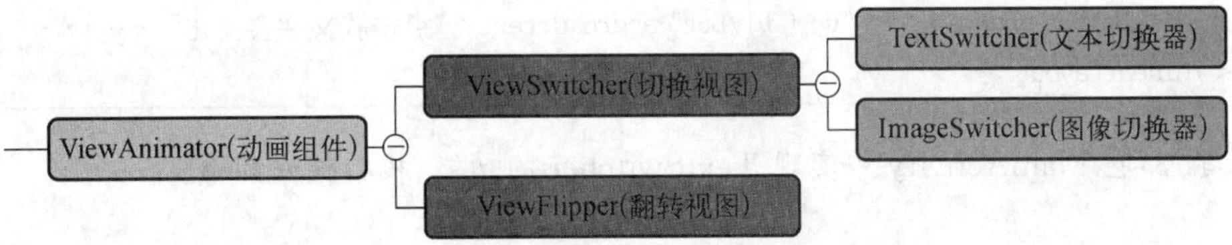

图 4 - 12　ViewAnimator 类组件结构

下面介绍基于 ViewAnimator 类的一些控件使用。

4.3.1 ViewSwitcher

ViewSwitcher 用于在视图之间切换,它使用视图工厂创建或添加视图,主要方法如下。

(1) addView (View child, int index, ViewGroup.LayoutParams params),其作用是产生添加一个带有指定布局参数的子视图,并将其加载在用户界面中。参数 child 为将被添加的子视图;参数 index 为子视图所在位置的值;参数 params 为子视图所设置的布局参数。异常的处理抛出 IllegalStateException。

(2) getNextView(),下一个要显示的视图。返回值 View 为要显示的下一个视图。

(3) reset (),其作用是重置和初始化视图切换器,隐藏所有存在的视图。

(4) setFactory (ViewSwitcher.ViewFactory factory),用来生成在切换器中切换的视图工厂类对象。参数 factory 用来生成切换器的视图工厂。也可以调用两次 addView (android. view. View, int, android. view. ViewGroup. LayoutParams) 来代替方法 setFactory。

1. TextSwitcher

TextSwitcher(文本切换器)的主要功能是使屏幕上的文本产生动画效果。与 ImageSwitcher 类一样,使用 TextSwitcher 类的时候依然需要通过 ViewSwitcher. ViewFactory 接口设置指定切换操作,并在 TextSwitcher 类中定义的常用方法。当调用 setText(CharSequence) 时,TextSwitcher 使用动画形式隐藏当前的文本并显示新的文本。下面通过一个案例介绍文本切换器的使用。

【微信扫码】
案例 4-5
相关文件

案例 4-5:TextSwitcher 的使用

在 Android Studio 中新建一个项目命名为 UI_Teach_TextSwitcher,实现可以自动切换的文本。

步骤 1:在资源文件 res/values/strings 中,新增几行用于显示的字符串,代码如下。

```
<string name = "text1">显示的文字</string>
```

步骤 2:打开 res/layout/activity_main.xml 编写布局文件,在垂直的线性布局中,放置一个文本切换器控件和一个按钮,核心代码如下。

```
<LinearLayout android:orientation = "vertical" …… >
    <TextSwitcher android:id = "@ + id /myTestSwitcher"……  />
    <Button android:id = "@ + id /but" android:text = "点击我文字会变化" ……  />
</LinearLayout>
```

步骤 3:在 MainActivity 中实现 TextSwitcher 的功能,核心代码如下。

```
public class MainActivity extends Activity {
    …… //声明控件
    @Override
    protected void onCreate(Bundle savedInstanceState) {
     …… //
     this.myTextSwitcher.setFactory(new ViewFactoryImpl()); //设置 TS 视图工厂
      this.myTextSwitcher.setInAnimation(AnimationUtils.loadAnimation(this, android.R.
anim.fade_in)); //设置 InAnimation 淡入方式
      this.myTextSwitcher.setOutAnimation(AnimationUtils.loadAnimation(this, android.R.
anim.fade_out)); //设置 InAnimation 淡出方式
     this.but.setOnClickListener(new OnClickListenerImpl()); //设置按钮的监听事件}
```

步骤 4：如果需要使用 TextSwitcher，先要设置 Switcher 的视图工厂，视图工厂的核心工作是负责产生具体显示内容的子视图。具体的实现代码如下。

```
public class ViewFactoryImpl implements ViewSwitcher.ViewFactory {
    @Override
    public View makeView() {
      TextView txt = new TextView(MainActivity.this); //子视图类型，文本使用 TextView
      txt.setBackgroundColor(0xFFFFFFFF); //设置 TextView 的背景颜色
      txt.setLayoutParams(new TextSwitcher.LayoutParams(
      LinearLayout.LayoutParams.MATCH_PARENT, LinearLayout.LayoutParams.MATCH_PARENT));
txt.setTextSize(20); //设置 TextView 的尺寸和字体大小
    return txt; //将设置好的子视图返回调用方法}}
```

步骤 5：Switcher 的内容不会自动切换，需要外部进行控制，这里使用按钮，通过按钮的点击事件来切换内容。按钮点击事件的实现代码如下。

```
public class OnClickListenerImpl implements View.OnClickListener {
 @Override
 public void onClick(View v) {
   if (index = = 4){index = 0;} //当 index 到达队尾时，循环回到队头
   switch(index){
   case 0:strshow = "当前的时间为:" + new SimpleDateFormat("yyyy - MM - dd HH:mm:ss.SSS").
format(new Date());break;
  case 1:strshow = (String)MainActivity.this.getResources().getString(R.string.text2);break;
  case 2:strshow = (String)MainActivity.this.getResources().getString(R.string.text3);break;
  case 3:strshow = (String)MainActivity.this.getResources().getString(R.string.text4);break;
  default:strshow = "……" ;break;}
         index++; //index 值来标记当前位置
       MainActivity.this.myTextSwitcher.setText(strshow); }}
```

运行程序，点击按钮，可以看见切换的文字，如图 4-13 所示。

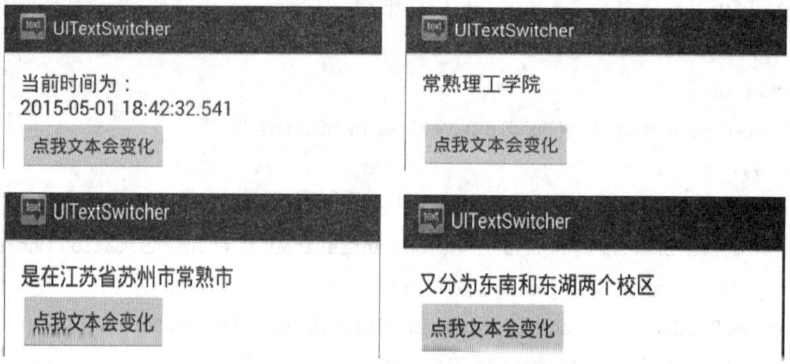

图 4-13 文本切换器程序的运行界面

2. ImageSwitcher

ImageSwitcher(图像切换器)的主要功能是完成图片的切换显示,例如用户在进行图片浏览的时候,可以通过按钮点击一张张图片切换显示。表 4-4 列出了一些常用的方法。

表 4-4 常用操作方法

方 法	描 述
public ImageSwitcher(Context context)	创建 ImageSwitcher 对象
public void setFactory(ViewSwitcher.ViewFactory factory)	设置 ViewFactory 对象,用于完成两个图片切换时 ViewSwitcher 的转换操作
public void setImageResource(int resid)	设置显示的图片资源 ID
public void setInAnimation(Animation inAnimation)	图片读取进 ImageSwitcher 时的动画效果
public void setOutAnimation(Animation outAnimation)	图片从 ImageSwitcher 要消失时的动画效果

在使用 ImageSwitcher 切换图片时,可以通过 Animation 指定切换图片时的动画显示效果,而 Animation 类的对象通过使用 AnimationUtils 类完成,主要方法如下。

public static Animation loadAnimation(Context context, int id)创建 Animation 对象。在使用 loadAnimation()方法创建 Animation 对象时需要指定操作的资源类型,这些类型可以直接从 android.R 类定义的常量中找出,使用以下两个资源常量:fade_in(进入时动画显示),fade_out(离开时动画显示)。

要想实现图片的切换功能,定义的 Activity 类还必须实现 ViewSwitcher.ViewFactory 接口,以指定切换视图的操作工厂,此接口定义如下。

```
public static interface ViewSwitcher.ViewFactory {
                                        public abstract View makeView(){};}
```

接口中只有一个抽象方法 makeView(),它的作用是创建一个新的 View,并将其加入 ViewSwitcher 之中,ImageSwitcher 的使用和 textSwitcher 类似,差别在于一个显示图像,一个显示文本。那么在视图工厂中,就要生成不同的子视图来放置显示的内容。

下面通过一个案例来学习如何使用 ImageSwitcher 实现图片的切换。

案例 4－6：ImageSwitcher 的使用

【微信扫码】
案例 4－6
相关文件

新建一个 Android 项目，命名为 UI_Teach_ImageSwitcher。

步骤 1：导入图片资源，放置一些图片到 res/drawable 中。

步骤 2：在 res/layout/activity_main.xml 中编写布局文件，在垂直的线性布局中，从上往下为放置按钮（前一张）、图像切换器、按钮（下一张），核心代码如下。

```
<LinearLayout android:orientation = "vertical" …… >
    <Button android:id = "@ + id /button1" android:text = "上一张" ……  />
    <ImageSwitcher android:id = "@ + id /imageswitcher"……  />
    <Button android:id = "@ + id /button2" android:text = "下一张"……  />
</LinearLayout>
```

步骤 3：修改 MainActivity 的代码。

```
public class MainActivity extends Activity {
    …… //声明程序中用到的组件和资源
  private int index = 0;
  @Override
  protected void onCreate(Bundle savedInstanceState) {
    …… //获得布局样式和控件
      this. imageSwitcher. setInAnimation ( AnimationUtils. loadAnimation ( this,  android. R.
anim. fade_in)); //设置切换器的淡入淡出方式
     this. imageSwitcher. setOutAnimation ( AnimationUtils. loadAnimation ( this,  android. R.
anim. fade_out));
    this. imageSwitcher. setFactory(new ViewSwitcher. ViewFactory() {
     @Override
     public View makeView() {
     ImageView imageView = new ImageView(MainActivity. this); //创建图像视图控件
     imageView. setScaleType( ImageView. ScaleType. FIT_CENTER); //设置视图纵横比
     imageView. setLayoutParams(new ImageSwitcher. LayoutParams(LinearLayout. LayoutParams.
WRAP_CONTENT,LinearLayout. LayoutParams. WRAP_CONTENT)); //设置视图参数
            return imageView; //返回这个图像视图} });
      imageSwitcher. setImageResource( imageId[ index]);}
```

步骤 4：下一张和上一张按钮的点击事件实现代码如下。

```
down. setOnClickListener(new View. OnClickListener() {
   @Override
   public void onClick(View v) { //当 index 到达最高值时，回归为初始值
     if (index < imageId. length - 1){ index ++ ;}else{ index = 0;}
     imageSwitcher. setImageResource( imageId[ index]);}}); }
     up. setOnClickListener(new View. OnClickListener() {
```

```
@Override
public void onClick(View v) { //当 index 等于 0 时,回归为最高值
  if (index>0){index--;}else{index = imageId.length - 1 }
  imageSwitcher.setImageResource(imageId[index]);  }   });
```

运行程序,点击上一张和下一张按钮,可以在不同的图片之间进行切换,如图 4 - 14 所示。

图 4 - 14 图像切换器的运行界面

4.3.2 ViewFlipper

ViewFlipper 组件继承于 ViewAnimator,属于比较简单的控件,不能自己加载切换子视图,需要手动操作。容器中可以添加多个子视图,但每次仅能显示一个子视图,同时可以使用动画控制切换效果,并可以设置间隔时间使子视图像幻灯片一样自动显示。主要方法如下。

(1) isAutoStart ():如果自动调用 startFlipping() 方法,则返回真。

(2) isFlipping ():如果已启动子视图定时切换,则返回真。

(3) setAutoStart (boolean autoStart):是否自动调用 startFlipping() 方法。

(4) setFlipInterval (int milliseconds):设置视图间切换的时间间隔,单位为毫秒。

(5) startFlipping ():开始在子视图间定时循环切换。

(6) stopFlipping ():停止切换。

案例 4 - 7:ViewFlipper 组件的使用

新建一个项目 UI_Teach_ViewFlipper,使用 ViewFlipper 实现一个可以自动播放的相册。

【微信扫码】案例 4 - 7 相关文件

步骤 1:在 res/layout/activity_main.xml 中编写布局文件。

```
<RelativeLayout android:layout_width = "match_parent" ...... >
  <ViewFlipper android:id = "@ + id /details" android:paddingTop = "50dp"
    android:flipInterval = "1000" android:persistentDrawingCache = "animation" ...... >
    <ImageView android:src = "@drawable /image1" ...... />
```

```
    ……//一共放置五个 ImageView 控件
  </ViewFlipper>
  <Button android:text="上下自动翻转" android:onClick="autoupdown" …… />
  ……//添加其他五个按钮,text + onClick 分别为上翻 + up、下翻 + down、左右自动 + auto、左
翻 + prev、右翻 + next"
</RelativeLayout>
```

在 ViewFlipper 组件中放置五个用来切换的子视图,子视图的翻转使用六个按钮实现。

步骤 2:在 MainActivity 中编写代码,实现切换的效果。

```
public class MainActivity extends Activity {
    private ViewFlipper viewFlipper;
    @Override
    protected void onCreate(Bundle savedInstanceState) {
        super.onCreate(savedInstanceState); setContentView(R.layout.activity_main);
        viewFlipper = (ViewFlipper) findViewById(R.id.details);}
    public void next(View source){
        viewFlipper.setInAnimation(this,android.R.anim.slide_in_left); //动画的滑入
        viewFlipper.setOutAnimation(this,android.R.anim.slide_out_right); //动画的滑出
        viewFlipper.showNext(); viewFlipper.stopFlipping();    }    }
```

方法中的参数 android.R.anim.slide_out_right 和 android.R.anim.slide_in_left 是用来控制滑动的方向。slide_in_left 表示从左边滑入,slide_out_right 表示从右边滑出。这两个 xml 文件是 Android 系统提供的,在左侧工程目录中将 Android 视图模式切换到 Project 模式,然后依次打开 External Libraries/ Android API xx Platform/res/anim,在里面找到如图 4-15 所示以 slide 开头的 XML 文件。打开 slide_in_left 文件,查看具体的实现方法,代码如下。

图 4-15　系统提供的默认 slide 文件

```
<set xmlns:android="http://schemas.android.com/apk/res/android">
  <translate android:fromXDelta="-50%p" android:toXDelta="0"   android:duration="
@android:integer/config_mediumAnimTime" />
  <alpha android:fromAlpha="0.0" android:toAlpha="1.0"   android:duration="@
android:integer/config_mediumAnimTime"  /></set>
```

translate 标签为定义转换动作;fromXDelta 为切换动作在 x 轴上的起始位置;toXDelta 为切换动作在 x 轴上的终止位置;alpha 为切换的透明度;fromAlpha="0.0"表示初始时为不显示;android:toAlpha="1.0"表示进入屏幕之后显示出来。

为了达到更多的滑动效果,在 res 上面右键单击,选择 Android resources directory,新建一个文件夹命名为 anim,接着在 anim 上右键单击,选择 Animation resources file,新建六个带有 set 标签的 xml。具体的实现如表 4-5 所示。

表 4-5　anim 动画效果文件列表

文件	作用	坐标轴	起始位置	坐标轴	终止位置
slide_in_right	右侧滑入	fromXDelta	100%p	toXDelta	0
slide_out_left	左侧滑出	fromXDelta	0	toXDelta	−100%p
slide_in_top	上侧滑入	fromYDelta	100%p	toYDelta	0
slide_out_bottom	下侧滑出	fromYDelta	0	toYDelta	−100%p
slide_in_bottom	下侧滑入	fromYDelta	−100%p	toYDelta	0
slide_out_top	上侧滑出	fromYDelta	0	toYDelta	100%p

每个按钮对应的动画文件关系如表 4-6 所示。

表 4-6　按钮设置

按钮名称	说明	滑入 xml 文件	滑出 xml 文件
up	上翻	slide_in_bottom	slide_out_top
down	下翻	slide_in_top	slide_out_bottom
atuoupdown	上下自动翻	slide_in_top	slide_out_bottom
prex	左翻	slide_in_right	slide_out_left
next	右翻	slide_in_left	slide_out_right
auto	左右自动翻	slide_in_left	slide_out_right

图 4-16　ViewFlipper 运行效果

根据表 4-6 中的对应关系，参考代码添加对应的按钮点击事件名称，并替换 setInAnimation 和 setOutAnimation 中对应的 anmin 的文件即可。运行结果，测试的效果如图 4-16 所示。

可以看到每当点击按钮时，按照预先的设置，可以实现左右或者上下的图片自动播放。

小　结

本章介绍了 Android 平台常用组件的使用，包括文本框和编辑框、按钮、单选和复选按钮、下拉列表、进度条类和切换类控件，本章所讲解的组件在实际开发中非常重要，基本上每个 Android 程序都会使用这些控件，因此初学者必须熟练掌握，为后面的学习做好铺垫。

课后练习

【微信扫码】
第 4 章课后练习

第 5 章

高级 UI 组件

5.1 Adapter 和 AdapterView

Adapter 为适配器。在 Android 系统中它是数据和视图之间的桥梁。它的作用是将数据绑定到 UI 界面上，并负责创建和显示每个项目子 View 和提供对下层数据的访问。当结构复杂的数据需要在各种视图中显示时，Adapter 是最合适有效的解决方案。

Adapter 类的子类有 ArrayAdapter，BaseAdapter，CursorAdapter，HeaderViewListAdapter，ListAdapter，ResourceCursorAdapter，SimpleAdapter，SimpleCursorAdapter，SpinnerAdapter，WrApperListAdapter 等。其中 ListAdapter，SimpleAdapter，ArrayAdapter 和 CursorAdapter 等继承自 BaseAdapter。Adapter 的抽象公共方法如表 5－1 所示。

表 5－1 Adapter 的抽象公共方法

方法名	描 述
int getCount ()	返回适配器的数据集中包含多少条目
Object getItem (int position)	获取数据集中指定位置的数据项目
long getItemId (int position)	取得列表中与指定位置的 ID
View getView (int position, View convertView, ViewGroup parent)	获得视图
int getItemViewType(int position)	获取由 getView 方法创建的指定位置的视图类型

其中参数 position 是指要从适配器中取得条目的位置。特别注意，上述 Adapter 的方法均为 abstract，在实现的子类中必须提供具体的方法。

AdapterView 是 Android 系统中用于以某种方式显示重复 View 对象的一类 View 总称。AdapterView 是 ViewGroup 的子类，它可以包含多个子 View。其特点是在一个 AdapterView 中会包含一系列的子 View 及用作显示的某个数据源。支持 Adapter 绑定的 UI 控件必须扩展 AdapterView 抽象类。使用这些 Adapter 类时，根据需要对 BaseAdapter 或者其他类型的 Adapter 进行扩展以满足功能上的需要。图 5－1 列出了 AdapterView 的控件继承关系。

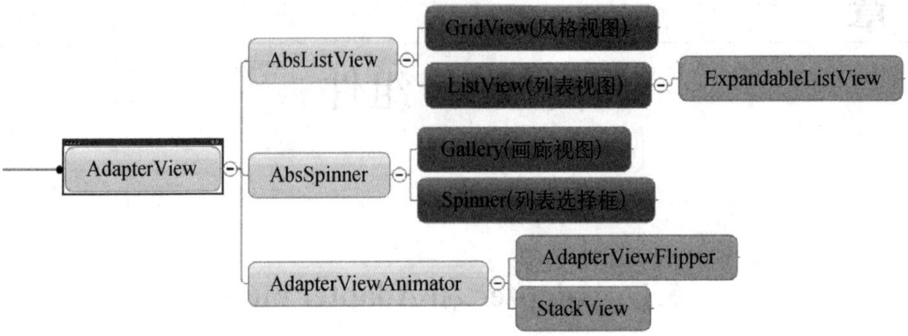

图 5-1 AdapterView 的结构

在程序中，将数据绑定到某个 Adapter 的子类中，再调用 View 的 setAdapter 方法将这个 Adapter 绑定到 AdapterView 的某个子类中，这样就可以在界面上显示数据。因为不同的 Adapter 子类搭配不同的 AdapterView 子类，下面将具体介绍二者之间的组合运用。

5.2 Spinner+ArrayAdapter

5.2.1 Spinner

Spinner 是每次只能选择所有项中一项的控件。装载的数据可通过它的 XML 属性 android:entries 对数组资源的引用获得。Spinner 继承了 AdapterView，因此下拉列表项将可以利用 Adapter 实现自动装配，而不需要在界面布局文件中一一声明。方法 setOnItemSelectedListener 为 Spinner 注册事件监听器。常用的方法如下。

(1) public SpinnerAdapter getAdapter ()：返回当前关联的适配器。

(2) public int getCount ()：返回与 AdapterView 关联适配器内的条目数量。

(3) public View getSelectedView ()：返回当前选中条目对应的视图。

(4) public void setAdapter (SpinnerAdapter adapter)：为 Spinner 提供后端数据。

5.2.2 ArrayAdapter

ArrayAdapter 的功能是将数组的每一个元素与布局资源中单独的一个 View 绑定。ArrayAdapter 操作的是数组或 List 数据。重载的构造函数如下。

```
ArrayAdapter (Context context, int resource, T[] objects)
ArrayAdapter (Context context, int resource, int textViewResourceId, List<T> objects)
ArrayAdapter<CharSequence> createFromResource (Context context, int textArrayResId, int
textViewResId)
```

方法中的参数 context 为应用程序的所在环境；参数 resource 为一个包含 TextView 的布局资源组件的 id，在实例化视图的时候使用；参数 objects 是和 View 组件绑定的数据；参数 textArrayResId 为数据源数组的 ID 值；参数 textViewResId 用于显示数据的视图资源 ID；参数 textViewResourceId 为和数据绑定的 View 组件的 ID。

下面使用 ArrayAdapter 结合 Spinner，介绍 Spinner 的数据加载方式。

案例 5-1：Spinner+ArrayAdapter 的组合程序

新建一个 Android 项目，命名为 UI_Teach_Spinner，使用三种方式加载下拉列表框数据。

【微信扫码】
案例 5-1
相关文件

步骤 1：在 res/values/strings.xml 文件中，新建两个字符串数组，分别命名为 city_labels 和 color_label，代码如下。

```
<string-array name="city_labels"><item>北京</item><item>上海</item><item>长沙</item><item>昆明</item><item>苏州</item></string-array>
<string-array name="color_label"><item>红色</item><item>蓝色</item><item>绿色</item><item>白色</item><item>橙色</item><item>黑色</item></string-array>
```

步骤 2：新建 res/layout/activity_main.xml 布局文件，放置三个 Spinner，代码如下。

```
<TextView android:id="@+id/info_city"  android:text="请选择你喜欢的城市" …… />
<Spinner android:id="@+id/mycity" android:entries="@array/city_labels" …… />
<TextView android:id="@+id/info_color" android:text="请选择你喜欢的颜色" …… />
<Spinner android:id="@+id/mycolor" />
<TextView android:id="@+id/info_edu" android:text="请选择你喜欢的学历"…… />
<Spinner android:id="@+id/myedu" />
```

布局界面效果如图 5-2 所示。编写之后先运行程序，第一个名称为 mycity 的 Spinner 控件，可以通过 android:entries 直接加载字符串数组 city_labels，所以"请选择你喜欢的城市"不需要编写代码即可得到数据，效果如图 5-3 所示。

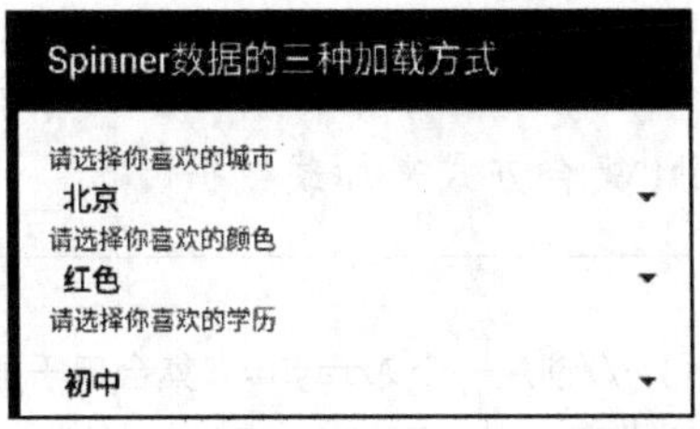

图 5-2　界面上的三个 Spinner 控件

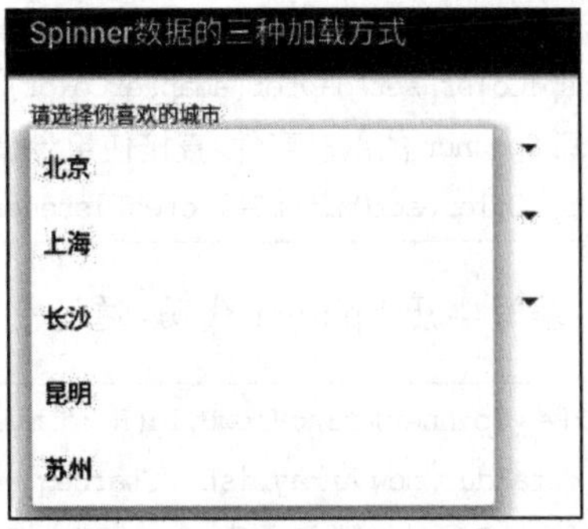

图 5-3　直接加载数据的 Spinner

步骤 4：在 MainActivity.java 中编写，实现获得所点击 Spinner 中的数据，代码如下。

```
public class MainActivity extends Activity {
    //声明两个 ArrayAdapter 适配器，用来给 mycolor 和 myedu 两个 Spinner 提供数据
    private ArrayAdapter<CharSequence> adapterColor = null;
    private ArrayAdapter<CharSequence> adapterEdu = null;
```

```
private List < CharSequence >   dataEdu = null; //声明一个 List 集合
//……声明需用的三个 Spinner 控件
@Override
protected void onCreate(Bundle savedInstanceState) {
 //……
spiCity = (Spinner) findViewById(R. id. mycity); //找到命名为 mycity 的 Spinner 控件
//mycity 的 Spinner 的点击事件
this. spiCity. setOnItemSelectedListener(new OnISListenerImpl());}
public class OnISListenerImpl implements AdapterView. OnItemSelectedListener {
@Override
public void onItemSelected(AdapterView <?> parent, View view, int position, long id){
    //获得当前选择的条目信息,并赋给字符串变量
    String valuestring = parent. getItemAtPosition(position). toString();
    Toast. makeText(MainActivity. this, "您选择的是:" + valuestring, Toast. LENGTH_SHORT).
show(); //使用 Toast 显示点击的内容}
    @Override
    public void onNothingSelected(AdapterView <?> arg0) { } }
```

步骤 5:实现通过适配器 ArrayAdapter 给名称为 mycolor 的控件加载数据。

```
spiColor = (Spinner) findViewById(R. id. mycolor); //找到声明的 Spinner 控件
piColor. setPrompt("请选择您喜欢的颜色");
this. adapterColor = ArrayAdapter. createFromResource(this, R. array. color_label, android. R.
layout. simple_spinner_item); //创建 ArrayAdapter 同时加载字符串数组和显示的视图资源
this. adapterColor. setDropDownViewResource ( android. R. layout. simple _ spinner _ dropdown _
item); //设置 Spinner 的下拉样式
this. spiColor. setAdapter(adapterColor); //将 Adapter 运用到控件上
//编写 Spinner 的点击事件,直接使用步骤 4 编写好的代码
this. spiColor. setOnItemSelectedListener(new OnItemSelectedListenerImpl());
```

步骤 6:编写学历 Spinner 代码,通过使用 List 集合方式来加载数据。

```
spiEud = (Spinner) findViewById(R. id. myedu);
this. dataEdu = new ArrayList < CharSequence >(); //创建一个 ArrayList 集合用于加载数据
this. dataEdu. add ( "初中"); this. dataEdu. add ( "高中"); this. dataEdu. add ( "大学"); this.
dataEdu. add("硕士"); this. dataEdu. add("博士"); //在 ArrayList 中添加数据
this. adapterEdu = new ArrayAdapter < CharSequence >(this, android. R. layout. simple_spinner
_dropdown_item, dataEdu); //使用 new 方法配置需要用的 ArrayAdapter
this. adapterEdu. setDropDownViewResource ( android. R. layout. simple _ spinner _ dropdown _
item); //设置 Spinner 的下拉样式
this. spiEud. setAdapter(adapterEdu); //将 Adapter 运用到控件上
//编写 Spinner 的点击事件,直接使用步骤 4 编写好的代码
this. spiEud. setOnItemSelectedListener(new OnItemSelectedListenerImpl());
```

步骤 5 和 6 编写完成之后，运行程序的效果如图 5-4 和图 5-5 所示。

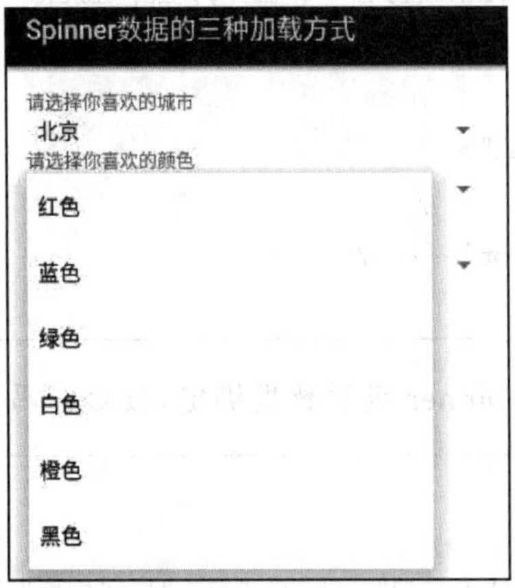

图 5-4　使用 ArrayAdapter 加载数据

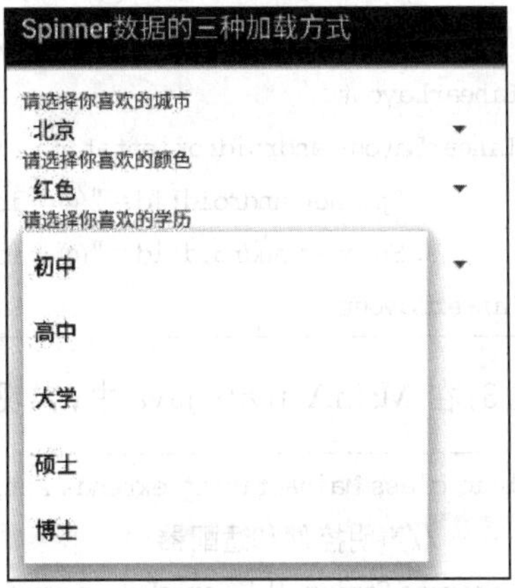

图 5-5　使用 List 加载数据

总结本案例的三种实现方式：① 城市为 city_date.xml 定义直接在 activity_main.xml 中加载。② 颜色为 color_date.xml 定义再通过 ArrayAdapter 加载。③ 学历为无 xml，用 ArrayList 中的数据再通过 ArrayAdapter 加载。

上面的案例实现了 Spinner 和数据的结合，但在实际应用中 Spinner 经常需要使用级联方式显示数据，就是当点击第一个 Spinner 的某个数据时，第二个 Spinner 加载对应的数据。下面介绍如何在 Spinner 中实现级联数据。

案例 5-2：下拉列表框级联数据的使用

【微信扫码】
案例 5-2
相关文件

新建一个 Android 项目，命名为 UI_Teach_SpinnerCascadeMenu。

步骤 1：新建一个 string-array 类，文件名为 city_data.xml，编写城市数据数组。

```
<string-array name="city_labels"><item>中国-北京</item>……</string-array>
<string-array name="region"><item>昆明</item><item>大理</item>……</string-array>

<string-array name="km"><item>滇池</item><item>石林</item>……</string-array>
<string-array name="dl"><item>鸡足山</item><item>苍山</item>……</string-array>

<string-array name="lj"><item>古城</item><item>玉龙雪山</item>……</string-array>
<string-array name="xsbn"><item>热带植物园</item>…… </string-array>
```

步骤 2：在 res/layout/activity_main.xml 布局文件中，放置两组 Spinner 控件，第一组使用数组实现级联，第二组使用资源文件实现级联，效果如图 5-6 所示，主要代码如下。

```
<LinearLayout android:orientation="horizontal">
        <Spinner android:id="@+id/city"  android:entries="@array/city_labels" ...... />
        <Spinner android:id="@+id/area" ......  />
</LinearLayout>
<LinearLayout android:orientation="horizontal">
          <Spinner android:id="@+id/region" ......  />
          <Spinner android:id="@+id/scenicspot" ......  />
</LinearLayout>
```

步骤 3：在 MainActivity.java 中，实现第一组 Spinner 级联数据绑定，核心代码如下。

```
public class MainActivity extends Activity {
  ...... //声明控件和适配器
    private String[][]  areaData = new String[][]{{"天安门","朝阳","海淀"},{"相城","常熟","
张家港","园区","木渎"}, {"昆明","大理","昭通","楚雄","玉溪","香格里拉"}}; //级联的数组
    @Override
    protected void onCreate(Bundle savedInstanceState) {
      ............
      city.setOnItemSelectedListene (new OnISListenerImpl()); }
  //当点击第1组第1个的Spinner控件时，将对应的数组信息加载到适配器上
  public class OnISListenerImpl implements AdapterView.OnItemSelectedListener{
    @Override
    public void onItemSelected(AdapterView<?> parent, View view, int position, long id){
      MainActivity.this.adapterArea = new ArrayAdapter<CharSequence>(MainActivity.this,
android.R.layout.simple_spinner_item, MainActivity.this.areaData[position]);
      MainActivity.this.area.setAdapter(adapterArea); //控件绑定 Adapter }
      @Override
      public void onNothingSelected(AdapterView<?> arg0) { }}
```

运行程序，第一个的 Spinner 的城市的数据直接通过资源文件获得，点击出现下拉列表后，点击其中的某项数据，通过 position，在字符串数组 areaData 中获得对应的数据，装在到对应的级联 Spinner 中，效果如图 5-6 和图 5-7 所示。

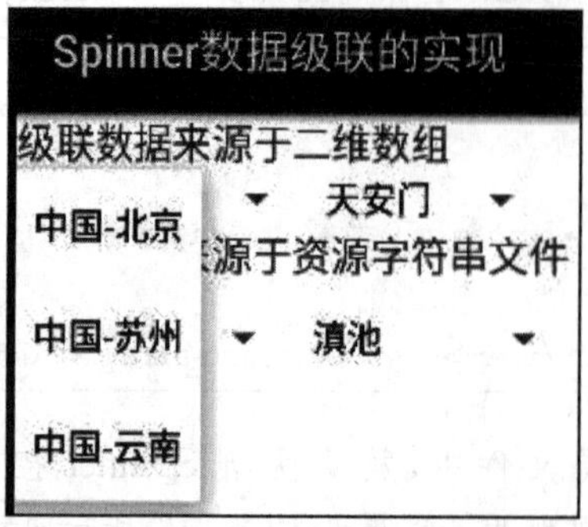

图 5-6　第一组 Spinner 数据显示

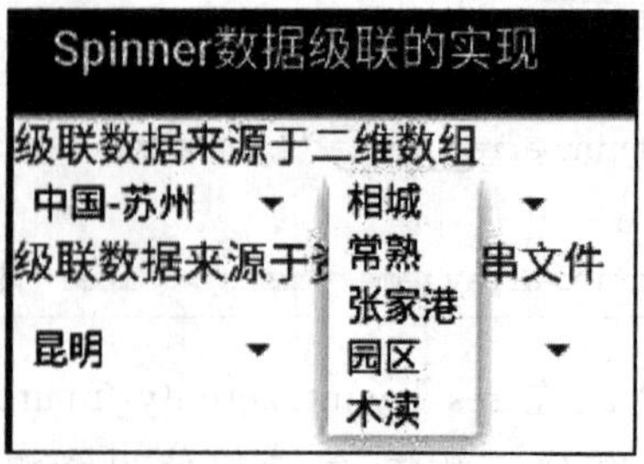

图 5-7　第一组 Spinner 级联的数据

步骤4:实现第二组 Spinner 级联数据,在 MainActivityonCreate()中新增如下代码。

```
    private int[]  scenicspotres = new int[]{R.array.km,R.array.dl,R.array.lj,R.array.xsbn};
  //创建适配器加载 city_data.xml 中 region 信息
  adapterregion = new ArrayAdapter < CharSequence >(this, android.R.layout.simple_spinner_
dropdown_item,this.getResources().getStringArray(R.array.region));
    region.setAdapter(adapterregion);
    this.region.setOnItemSelectedListener(new OnRegionItemSelectedListenerImpl());}
  public class OnRegionItemSelectedListenerImpl implements AdapterView.OnItemSelectedListener {
    @Override
  public void onItemSelected(AdapterView<?> parent,View view,int position,long id){
        adapterscenicspot = new ArrayAdapter < CharSequence >(MainActivity.this, android.R.layout.
simple_spinner_item, MainActivity.this.getResources().getStringArray(scenicspotres
[position])); //建立 regionSpinner 和 scenicSpinner 的对应关系
  MainActivity.this.scenicspot.setAdapter(MainActivity.this.adapterscenicspot);}
  @Override
  public void onNothingSelected(AdapterView<?> arg0) { }  }}
```

当获得第一组 regionSpinner,点击位置后,找到 XML 中对应的数据里面,加载到第二组第二个 Spinner 中,如图5-8和5-9所示显示。

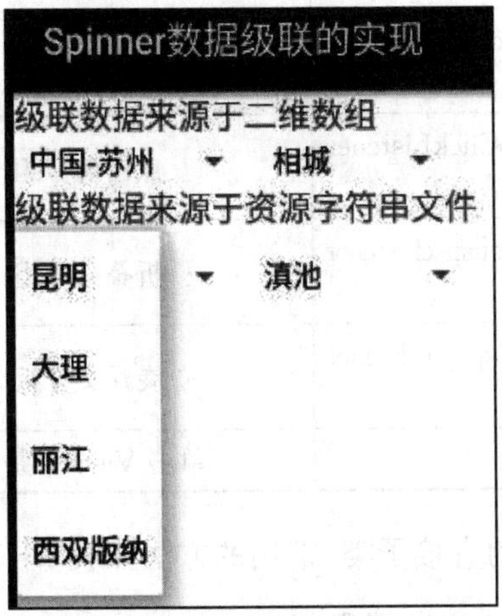

图5-8　第二组第一个 Spinner

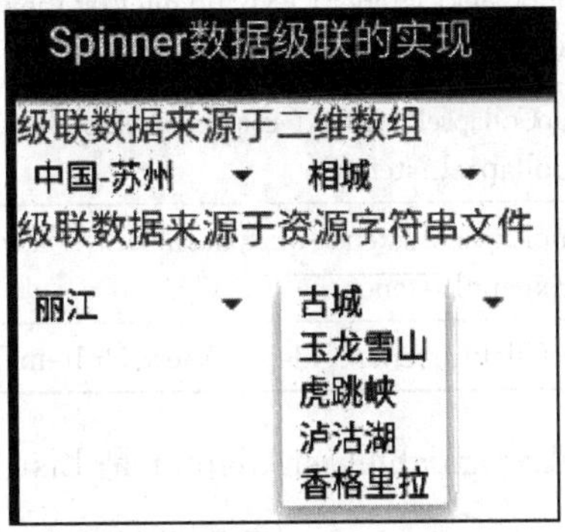

图5-9　第二组第二个 Spinner

5.3　ListView+ListAdapter

5.3.1　ListAdapter

ListAdapter 是 ListView 和(列表形式)数据之间的桥梁,当数据加载到 ListAdapter 中,就可以提供对数据的访问。ListAdapter 直接继承 Adapter,且需要实现 Adapter 的抽象方法。

5.3.2 ListView

手机的屏幕空间有限，一个屏幕只能显示有限的数据。当需要显示的内容比较多的时候，就借助 ListView 来实现，每一屏 ListView 都显示当前的部分数据，用户可以通过手指的上下滑动，将屏幕外的数据加载到屏幕上显示，屏幕上原来的数据将会滚动出屏幕。手机上的联系人列表、微信联系人和朋友圈信息都是使用 ListView 来实现。ListView 支持单选或多选等模式，可以对用户的单击事件进行处理。ListView 的点击方法使用 setOnItemClickListener()监听器，里面调用 onItemClick()方法获得点击的任何一个子项。同时调用 setTextFilterEnabled(true)允许应用根据用户的输入过滤 ListView 的列表项。

5.3.3 ExpandableListView＋BaseExpandableListAdapter

ListView＋ListAdapter 只能实现一级分组，在实际运用中，经常需要实现多级分组。它们的直接子类 ExpandableListView＋BaseExpandableListAdapter 可以实现。二者的结合可以实现多级分组的数据在垂直滚动的列表中显示，可以单独展开显示每个子级的组，展开列表在每个项目旁边显示一个指示器，以显示项目的当前状态(状态通常是展开组、折叠组)。ExpandableListView 是 ListView 的直接子类，常用方法如表 5－2 所示。

表 5－2 ExpandableListView 常用方法

方法名	描述
setOnChildClickListener(ExpandableListView.OnChildClickListener onChildClickListener)	单击子分组
setOnGroupClickListener(ExpandableListView.OnGroupClickListener onGroupClickListener)	点击父分组
setOnGroupCollapseListener(ExpandableListView.OnGroupCollapseListener onGroupCollapseListener)	折叠父分组
setOnGroupExpandListener(ExpandableListView.OnGroupExpandListener onGroupExpandListener)	展开父分组
setOnItemClickListener(AdapterView.OnItemClickListener l)	单击 View 中条目

BaseExpandableListAdapter 是 ListAdapter 的直接子类，常用的方法如表 5－3 所示。

表 5－3 BaseExpandableListAdapter 常用方法

方法名	描述
Object getGroup(int groupPosition)	由父分组位置获得父分组
int getGroupCount()	获得父分组的数量
long getGroupId(int groupPosition)	获得父分组的 ID 值
void onGroupCollapsed(int groupPosition)	注册父分组的折叠状态
void onGroupExpanded(int groupPosition)	注册父分组的展开状态
Object getChild(int groupPosition, int childPosition)	获得子分组
long getChildId(int groupPosition, int childPosition)	获得子分组的 ID 值

续表

方法名	描述
int getChildrenCount(int groupPosition)	获得子分组的数量
View getGroupView (int groupPosition, boolean isExpanded, View convertView, ViewGroup parent)	获得父分组的视图类型
View getChildView (int groupPosition, int childPosition, boolean isLastChild, View convertView, ViewGroup parent)	获得子分组的视图类型

下面通过案例来学习两者的使用。

案例 5-3:ListView 和 ListAdapter 的使用

新建一个项目，命名为 Adapter _ Teach _ ListAdapter _ ListView _ Expandable。

【微信扫码】案例 5-3 相关文件

在本案例中，将多组数组通过 ListAdapter 加载到 ListView 中，可以实现分数据的伸展和收缩功能。ListAdapter 和 ListView 两者本身功能一般，我们使用功能更强的子类 ExpandableListView 和 BaseExpandableListAdapter。

步骤 1:打开 res/layout/activity_main.xml 文件，放置一个 ExpandableListView，代码如下。

```
< ExpandableListView android: id = "@ + id /elistview" android: layout_width = "match_parent"
        android: layout_height = "wrap_content"  />
```

步骤 2:编写适配器的代码，在 src 中新建一个 java 类，命名为 MyExpandableListAdapter，继承 BaseExpandableListAdapter。

```
public class MyExpandableListAdapter extends BaseExpandableListAdapter {
  private String[ ] groups = new String[ ]{"计算机学院","机械学院",……}; //大分组数据
    private String[ ][ ] children = new String[ ][ ]{{"网络工程","计算机科学","软件工程", "
数字媒体" },{"机械制造","机械原理" },{"日语","英语","德语"},……}}; //子分组数据
    private Context context  = null ;
    public MyExpandableListAdapter(Context context){this. context = context;}
    private TextView buildTextView(){……return textView;} //设置一级分组文本样式
    private TextView buildChildTextView(){……return textView;} //设置二级分组文本样式
    @Override
     //实现 BaseExpandableListAdapter 类的抽象方法,每个方法之前都有@Override 装饰
     public Object getChild ( int groupPosition, int childPosition) { return this. children
[groupPosition][childPosition];}
    public long getChildId( int groupPosition,  int childPosition) {return childPosition;}

     public View getGroupView ( int groupPosition,  boolean isExpanded, View convertView,
ViewGroup parent) { …… return textView;}
    public View getChildView(int groupPosition, int childPosition, boolean isLastChild, View
convertView, ViewGroup parent) { …… return textView;}
```

```
public Object getGroup(int groupPosition) {return this.groups[groupPosition];}
public int getGroupCount() {return this.groups.length;}
public long getGroupId(int groupPosition) {return groupPosition;}
public boolean hasStableIds() {return true;}
public boolean isChildSelectable(int groupPosition, int childPosition){return true;}
public int getChildrenCount(int groupPosition) {return this.children[groupPosition].length;}}
```

步骤 3:在 MainActivity 的 onCreate 中添加代码实现对大分组和小子项进行设置,代码如下:

```
super.registerForContextMenu(this.elistview); //ListView 注册
this.elistview.setOnGroupClickListener(new OnGroupClickListenerImpl()); //点击大分组
this.elistview.setOnGroupExpandListener(new OnGroupExpandListenerImpl()); //展开
this.elistview.setOnGroupCollapseListener(new OnGroupCollapseListenerImpl()); //收缩
this.elistview.setOnChildClickListener(new OnChildClickListenerImpl()); //点击子项
```

步骤 4:具体实现上述几个方法。

```
private class OnChildClickListenerImpl implements ExpandableListView.OnChildClickListener {
    @Override
    public boolean onChildClick(ExpandableListView parent, View v, int groupPosition, int
childPosition, long id) {…… return false;}}
    …… //省略其他四个方法代码
```

每个 setXXXListener 类都要实现类中的 onXXXClick()方法,初次编写会觉得比较烦琐,但是因为基本都是固定模式,很容易掌握。运行程序,父分组全部折叠和子分组全部展开的效果如图 5-10 和 5-11 所示。

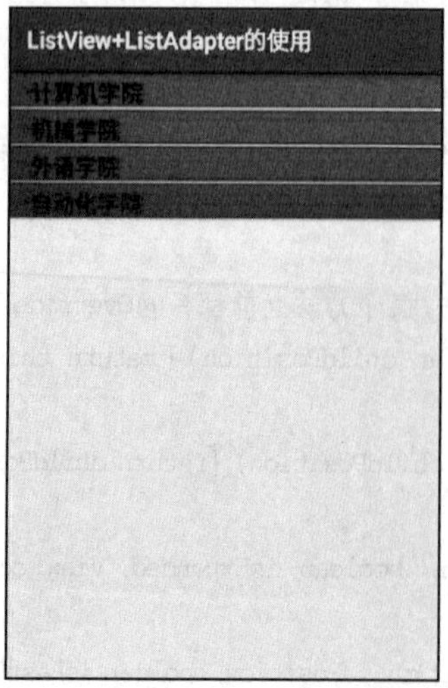

图 5-10　父分组全部折叠效果

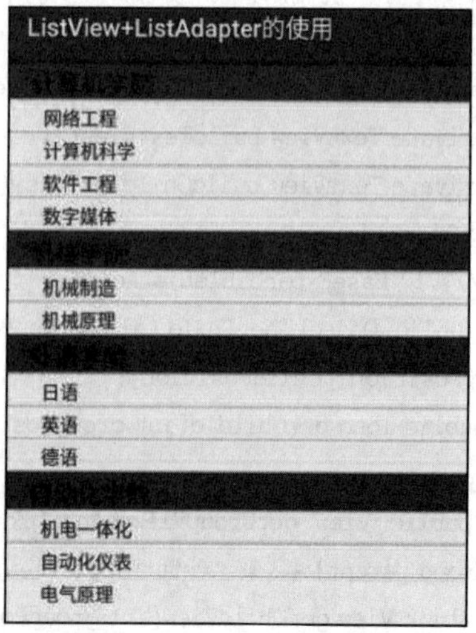

图 5-11　子分组全部展开效果

5.4　RecyclerView＋Recycler.Adapter

Google 的工程师在 support - v7 包中引入了一个全新的列表控件 RecyclerView，它在使用上比 ListView 和 GridView 更为灵活。它对应的适配器为 Recycler.Adapter，下面进行详细介绍。

5.4.1　RecyclerView

RecyclerView 是反复循环的视图，是一个用来显示庞大数据集的视图组件。通过保持有限数量的视图来循环显示大量的数据集。

使用 RecyclerView 控件，首先需要创建一个继承 RecyclerView.Adapter 的适配器，它负责将数据与每一个 Item 的界面进行绑定。其次需要一个 LayoutManager 布局管理器，它用来确定每一个 Item 如何进行排列、何时进行显示或者隐藏。当每一个 View 被创建或者回收时，LayoutManager 负责向 Adapter 请求用新的数据替换旧的数据。RecyclerView 提供了三种内置的布局管理器：① LinearLayoutManager：线性布局，可以使用水平或者垂直方式显示。② GridLayoutManager：网格布局，类似于 GridView。③ StaggeredGridLayoutManager：流式布局，比如瀑布效果。

5.4.2　RecyclerView.Adapter

RecyclerView.Adapter 将绑定的数据提供到 RecyclerView 视图中显示，主要方法如下。

(1) void bindViewHolder(VH holder, int position)：在给定的 position 中，更新 RecyclerView.ViewHolder 中的内容。内部调用 onBindViewHolder(VH,int)方法。

(2) void onBindViewHolder(VH holder,int position)：当每个子项滚动到屏幕中时，对每个 RecyclerView 子项的数据进行更新。参数 position 为当前子项的位置。

案例 5 - 4：反复循环视图 RecyclerView 的使用

【微信扫码】
案例 5 - 4
相关文件

新建一个项目，命名为 Adapter _ Teach _ RecyclerViewAdapter _ RecyclerView。

步骤 1：添加依赖。使用 RecyclerView 需要在 app/build.gradle 中添加依赖，导入的版本和 Appcompat 的版本保存一致。例如 'com.android.support：Appcompat - v7：28.0.0'，那么 recyclerview 的写法为 'com.android.support：recyclerview - v7：28.0.0'。

步骤 2：导入资源(图片和字符串数组)。导入程序需要的图片资源到 res/drawable 中，并在 res/values/strings 中新增图片对应的字符串数组。

步骤 3：编写布局文件(两个布局文件)。第一个是小布局文件 fruit_item.xml，在垂直的线性布局里面放置两个 TextView 和一个 ImageView，用来显示数据的小 View，代码如下。

```
< LinearLayout android:orientation = "horizontal"  ……>
    < TextView android:id = "@ + id /fruit_name" …… />
    < ImageView android:id = "@ + id /fruit_image" …… />
    < TextView android:id = "@ + id /fruit_desc" …… />
< /LinearLayout >
```

第二个是大的布局文件 activity_main.xml,放置一个 RecyclerView 控件。

```
< LinearLayout android:layout_width = "match_parent" …… >
   < android.support.v7.widget.RecyclerView android:id = "@ + id /recycler_view" ……  />
< /LinearLayout >
```

因为 RecyclerView 并不在默认的 SDK 中,所以导入的控件名称需要写入完整的路径。

步骤 4:在 src 中新建一个类命名为 Fruit 的实体类,类中有三个字段 name(水果名称)、imageId(水果图片)、desc(水果说明)和 get 属性。因为不需要写入信息,没有 set 属性。

步骤 5:新建一个继承 RecyclerView.Adapter 适配器的类 FruitAdapter,泛型为 FruitAViewHolder,包括控件的缓存类 FruitAViewHolder,代码的核心框架如下。

```
public class FruitAdapter extends Adapter < FruitAdapter.ViewHolder > {
    private List < Fruit > myFruitList; //定义一个 List 用于存放类型为 Fruit 的数据
    public FruitAdapter(List < Fruit > myFruitList){this.myFruitList = myFruitList;}
     //使用 LayoutInflater 加载小布局文件,传入 ViewHolder,返回实例化的 ViewHolder
    @Override
    public ViewHolder onCreateViewHolder(ViewGroup parent, int viewType){ …… }
     //通过 position 获得当前项实例,并对子 View 进行数据绑定
    @Override
    public void onBindViewHolder(ViewHolder holder, int position) { ……}
    @Override
    public int getItemCount() {return myFruitList.size(); } //获得一共有多少项数据
     //保持每个 Item 的界面数据元素复用 View,对其进行优化,避免频繁调用 findViewById 方法
和创建过多的 View
    static class ViewHolder extends RecyclerView.ViewHolder { //自定义 ViewHolder
        ImageView fruitImage;TextView fruitName;TextView fruitDesc;
        public ViewHolder(View view) { …… }}
```

步骤 6:修改 MainActivity 代码。

```
public class MainActivity extends AppCompatActivity {
    private List < Fruit > fruitList = new ArrayList <>(); //将数据放入到 List 中
    …… //声明控件和用于存放水果的字符信息、图片信息和描述信息的数组
    @Override
    protected void onCreate(Bundle savedInstanceState) {
        …………
```

```
LinearLayoutManager layoutManager = new LinearLayoutManager(this);
recyclerView.setLayoutManager(layoutManager); //RecyclerView 由 LinearLayoutManager 管理
FruitAdapter adapter = new FruitAdapter(fruitList); //设置和绑定适配器
recyclerView.setAdapter(adapter);}    }}
```

运行程序,效果如图5-12所示。

步骤7:实现单击事件。在ListView中要实现对数据子项的单击事件,只要注册setOnItemClickListener监听方法即可。但是在RecyclerView中,却发现没有这样的方法。那么RecyclerView为什么没有这样的事件监听?怎么实现对子项的单击事件呢?

在数据View中,一条数据可能由几个部分组成,比如编号+文字+图片,ListView的单击事件只能对数据的整条进行点击处理,不能对其中的某个部分单独设置点击事件,这样显得非常不方便,因此RecyclerView直接放弃了ListView的子项目单击处理方法,采用由具体的View去注册的方法。具体实现如下。

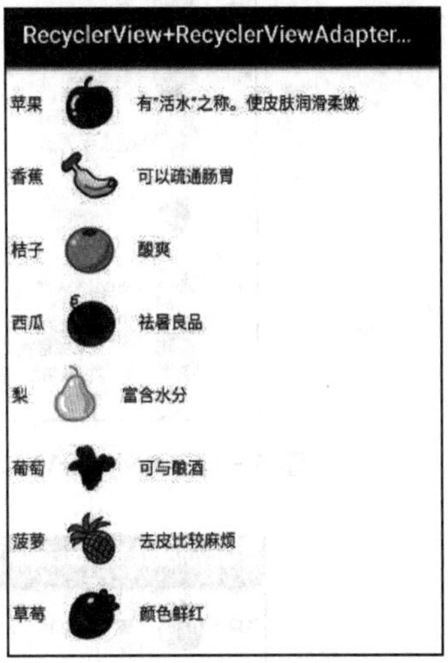

图5-12 RecyclerView的运行效果

在FruitAdapter类的onCreateViewHolder方法中新增如下代码。

```
//实现对数据子项的整体的点击事件,运行效果如图5-12
holder.fruitView.setOnClickListener(new View.OnClickListener(){
    @Override
    public void onClick(View view) {
        int position = holder. getAdapterPosition ( ); Fruit fruit = myFruitList. get
(position);
        Toast.makeText(view.getContext(),"你点击的\n 水果是:" + fruit.getName() + ",\n 功
效说明为:" + fruit.getDesc(),Toast.LENGTH_SHORT).show();}});
holder.fruitDesc.setOnClickListener(new View.OnClickListener(){……}); //子项名称单击事件
holder.fruitImage.setOnClickListener(new View.OnClickListener(){……}; //子项描述单击事件
```

运行程序,分别点击Item的整体、Item的名称和Item的描述,显示效果分别如图5-13、图5-14和图5-15所示。

步骤8:实现横向滚动和瀑布效果。ListView的布局是由自身进行管理,当需要进行改变时,就显得比较呆板,难以实现。RecyclerView使用LayoutManager来管理,制定了一套可扩展的布局排列接口,子类只需要按照接口规范来调用就能实现,大大减少了编码量。

在MainActivity中新增如下代码,就可以实现这个效果,如图5-16所示。

```
layoutManager.setOrientation(LinearLayoutManager.HORIZONTAL);
StaggeredGridLayoutManager layoutManager = new StaggeredGridLayoutManager(3,StaggeredGrid
LayoutManager.VERTICAL);
```

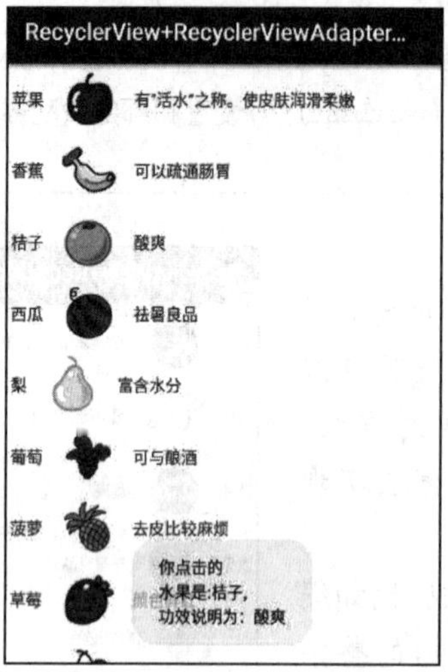

图 5－13 RecyclerView 子项的整体单击

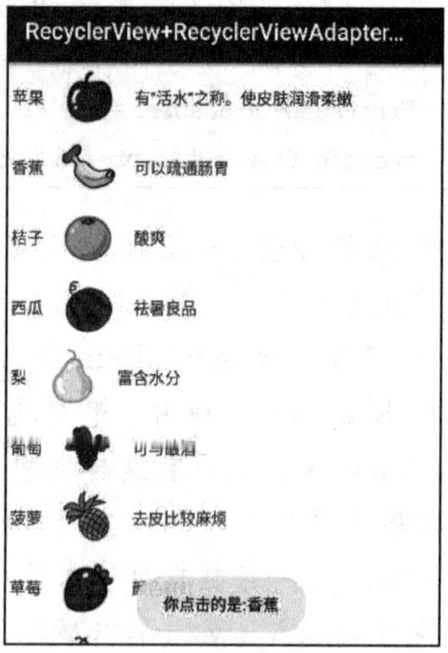

图 5－14 子项中点击名称的效果

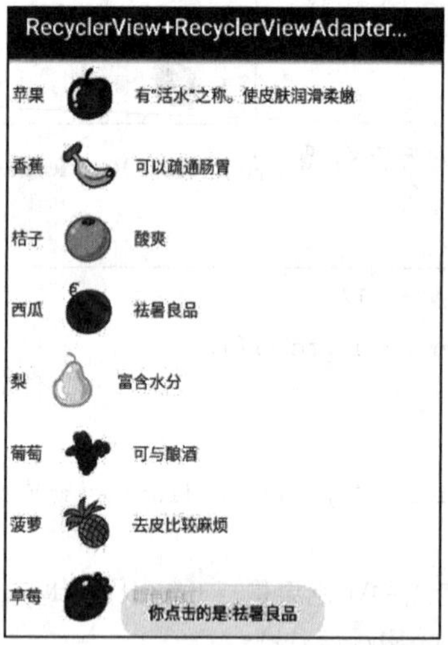

图 5－15 子项中单击描述的效果

图 5－16 RecyclerView 的瀑布效果

5.5 GridView＋SimpleAdapter

5.5.1 GridView

GridView 和 ListView 都是比较常用的布局控件，与 ListView 组件不同的是，它可以以二维网格的方式展示一系列数据。对于 Gridview 中的每个项目，还需要定义一个专门的子布局，用来显示每一条数据。

5.5.2 SimpleAdapter 组合

SimpleAdapter 为简单适配器，可以将静态数据映射到 XML 文件中定义好的视图，它操作的是 List 数据。可以将 Map 的 ArrayList 指定为用于列表的数据，ArrayList 中的每一项对应列表中的一行，适配器的构造函数如下。

```
public SimpleAdapter (Context context, List <? extends Map < String, ?>> data, int resource,
String[ ] from, int[ ] to)
```

其中的参数：① context 为上下文。② Map 包含用于一行的数据，可以指定 XML 文件定义了用于显示行的视图，通过 Map 的关键字映射到指定的视图。data 是绑定到 View 组件的数据，数据类型为 Map < String，? >。③ resource 表示布局文件的 ID。④ from 为 data 中的 key 值数字，to 是和数据绑定的 View 组件的 ID 数组，数据绑定时会按照 from 中 key 值和 to 中 ID 值的对应顺序，将 data 中的 key 所对应值绑定到相应的 View 组件中。

SimpleAdapter 绑定数据到视图后，会确认 SimpleAdapter. ViewBinder 是否有效，是则调用 setViewValue(android.view.View，Object，String) 方法。如果没有数据绑定发生，将会抛出 IllegalStateException 异常。

【微信扫码】
案例 5－5
相关文件

案例 5－5：GridView 结合 SimpleAdapter 的运用

新建一个 Android 项目，命名为 Adapter _ Teach _ SimpleAdapter _ GridView。

步骤 1：在 res 中新建一个字符串资源文件，命名为 imageTitle.xml，代码如下。

```
< string-array name = "imagetitle">
        < item >花开富贵< /item >< item >海天一色< /item > ……
< /string-array >
```

再新建一个字符串资源文件，命名为 imagecontent.xml，代码如下。

```
< string-array name = "imagecontent">
        < item >美丽的花朵< /item >< item >融入大自然的颜色< /item >……
< /string - array>
```

步骤 2：编写布局文件(两个布局文件)。

(1) 子布局 item.xml，数据项的每条包含三个内容，水果名称、水果图片和水果说明，因此放置两个文本框和一个 ImageView，用来显示数据的小 View，代码如下。

```
< TextView android: id = "@ + id /title" ……  />
< ImageView android: id = "@ + id /image" ……  />
< TextView android: id = "@ + id /content" ……   />
```

(2) 主布局 activity_main.xml，放置 GridView 控件，设置每行包含三个数据项，代码如下。

```
< GridView android: id = "@ + id /gridview1"   android:numColumns = "3" …… >< /GridView >
```

步骤 3：修改 MainActivity 代码。

```
public class MainActivity extends Activity {
    private int[] imageId = new int[]{R.drawable.img01,……}; //加载图片资源
 @Override
 protected void onCreate(Bundle savedInstanceState) {
    …… //将文字(两组)加载到 String 数组中
    String[] title = super.getResources().getStringArray(R.array.imagetitle);
    String[] content = super.getResources().getStringArray(R.array.imagecontent);
     //创建一个 List 集合并将图片和文字添加到集合中
    List < Map < String, Object >> listitems = new ArrayList < Map < String, Object >>();
    for (int i = 0; i < imageId.length; i ++ ) {  //装配 SimpleAdapter 并将适配器与 GridView
关联
     Map < String, Object > map = new HashMap < String, Object >();
     map.put("image", imageId[i]);map.put("title", title[i]);map.put("content", content[i]);
     listitems.add(map);}
     SimpleAdapter adapter = new SimpleAdapter(this, listitems, R.layout.items, new String[]
{"title", "image", "content"}, new int[]{R.id.title, R.id.image, R.id.content});
     gridview.setAdapter(adapter);     }}
```

运行程序,效果如图 5-17 所示。

图 5-17　GridView+SimpleAdapter 运行效果

5.6　BaseAdapter+Gallery

5.6.1　BaseAdapter

BaseAdapter 为基础适配器,它包含较多的方法,具有较高的灵活性。它是一个抽象类,如果要使用它,在继承的实体类中,开发人员要实现以下四个方法。

(1) public int getCount():返回后台一共有多少数据。

(2) public Object getItem(int position):返回指定位置的数据对象。

(3) public long getItemId(int position):返回指定位置的数据对象 ID。

(4) public View getView(int position, View convertView, ViewGroup parent) :返回一个加载了数据的 View,以便其他组件的调用。

5.6.2 Gallery

Gallery 用来横向展示一系列图像。Gallery 组件开发的核心,就是创建一个能够返回 ImageView 的 Adapter。这个 Adapter 一般是从 Baseadapter 继承,开发人员需要实现 getView、getCount 等几个重要的回调方法,最后将 Adapter 与 Gallery 绑定即可。

案例 5-6:BaseAdapter 和 Gallery 的使用

新建一个 Android 项目,命名为 Adapter_Teach_BaseAdapter_Gallery。

步骤 1:在 res/layout 中,修改 activity_main.xml 文件,在布局文件中首先放置一个 ImageSwitcher 控件,用于图像切换。然后在界面底部放置画廊 Gallery。主要代码如下。

【微信扫码】
案例 5-6
相关文件

```
<LinearLayout android:id="@+id/llayout" …… >
    <ImageSwitcher android:id="@+id/imageSwitcher"……  />
    <Gallery android:id="@+id/gallaery1" android:unselectedAlpha="0.6"…… />
</LinearLayout>
```

步骤 2:编写主程序文件,代码如下。

```
public class MainActivity extends Activity {
    ……//加载图片资源,声明程序中用到的组件
    @Override
    protected void onCreate(Bundle savedInstanceState) {
      ……//设置 ImageSwitcher 的图片的切换效果,请参照案例 4-6 的步骤 3
     imageSwitcher.setFactory(new ViewSwitcher.ViewFactory() {
       @Override
       public View makeView() {……//设置 ImageSwitcher 的视图工厂}          });
```

步骤 3:配置 BaseAdapter,在 MainActivity.java 文件,新增代码如下。

```
BaseAdapter adapter = new BaseAdapter(){
  ……//重写三个方法,int getCount()获得画廊上的图片的数量、Object getItem(int position)
获得画廊上指定位置的图片对象、long getItemId(int position) {return position;}获得画廊上
的指定图片的位置 ID 值
  @Override
  //返回指定位置上配置好的图片
```

```
public View getView(int position,View convertView,ViewGroup parent){
        ImageView imageView = null;
      //首先确定返回的 View 类型 ImageView,接着判断是否存在这样的 view,如果没有就进行
初始化配置,如果有就将 convertView 转化为 imageView 返回包含图片的 View }
        //将图片装配 Adapter 到组件上,并设置图片大小为实际大小一半
        gallery.setAdapter(adapter);gallery.setSelection(imageId.length /2);
        gallery.setOnItemSelectedListener(new AdapterView.OnItemSelectedListener(){
      @Override
      //设置画廊的点击事件,当某张图片被选中的时候,加载到 ImageSwitcher 中
      public void onItemSelected(AdapterView<?> parent, View view, int position, long id)
{ imageSwitcher.setImageResource(imageId[position]); }
        @Override
        public void onNothingSelected(AdapterView<?> arg0){ }});}}
```

图 5-18　BaseAdapter 和 Gallery 的使用效果

运行程序可以看到,底部的画廊加载图片,可以自由地左右滑动。滑动到中间的图片,淡入淡出效果变得清晰,同时会显示在上面的 ImageSwitcher 中。效果如图 5-18 所示。

小　结

AdapterView 是以某种特定方式显示重复 View 对象的一类 View 总称,典型代表有 ListView,GridView 和 GalleryView。Adapter 是将数据绑定到 UI 界面上的桥接类,它是 AdapterView 和底层数据之间链接的桥梁,它提供了对数据项的访问机制,同时还可以为数据集内的每一个数据提供创建好的视图,比较常用的有 ArrayAdapter,SimpleAdapter,BaseAdapter 等。本章内容非常重要,可以让读者掌握如何将复杂数据加载到控件上,请多练习、多实践。

课后练习

【微信扫码】
第 5 章课后练习

第 6 章

Activity 和 Intent

6.1 Activity

Activity 是一个独立的,可以与用户交互的 Android 应用组件。从功能上来说,它作为应用程序的界面框架,负责动态加载各种用户界面,实现底层的消息传递等。从逻辑上来说,Activity 是 Android 应用的组成部分,一个 Android 应用可以包含多个 Activity。

Android 系统将 Activity 等应用组件作为系统资源统一管理,对于用户请求,Android 系统像搭积木一样,灵活调用各个不同应用中的组件来完成指定的任务,这种方式大大提高组件的重用度,也是适应 Android 设备资源限制条件下的一种合理的设计思想。注意:虽然 Activity 是 Android 应用的组成部分,但 Activity 是可以独立于 Android 而运行的。

6.1.1 Activity 生命周期

在不同状态的 Activity 之间的生命周期方法的调用,只是正常状态下的一种过程。如果 Activity 在运行过程中发生意外,Activity 将被立即强制退出,而不会有机会执行其他生命周期方法。这一点开发人员要特别注意。

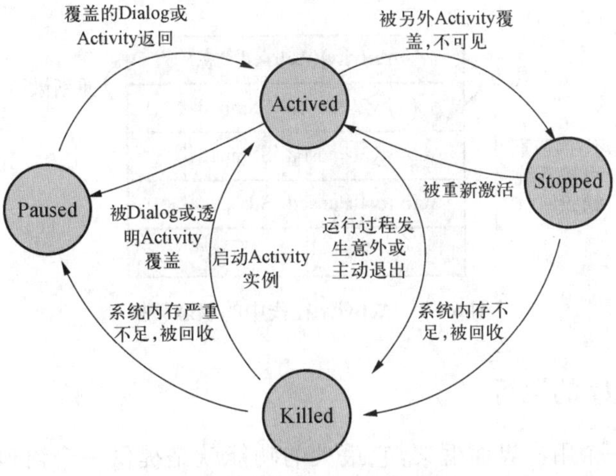

图 6-1 Activity 生命周期

当 Activity 运行的设备环境发生变化,如设备从横屏切换到竖屏,Android 会将当前的 Activity 销毁,并重新创建一个新的 Activity,而且会导致 onCreate 方法被调用。

(1) Actived:运行状态。此时它处于和用户交互的激活状态,界面显示在当前屏幕上。

(2) Paused:暂停状态。此时 Activity 被另一个透明或者 Dialog 样式的 Activity 覆盖,

它对用户仍然部分或全部可见,但已经失去了焦点,不可与用户交互。而 Android 会保持 Activity 的相关信息包括成员变量、与窗口管理器的连接等。

(3) Stopped:停止状态。当 Activity 被另外一个 Activity 覆盖、失去焦点并不可见时,则处于 Stopped 状态。在此状态下,Android 会保持 Activity 的相关信息包括状态和成员变量等。

(4) Killed:消亡状态。Activity 被系统杀死回收或者没有被启动时,则处于 Killed 状态。

处于 Paused 或 Stopped 状态下的 Activity,Android 会保存它的内部状态变量。但当系统资源紧张时,Activity 很有可能被 Android 杀死,以便回收系统资源。在这种情况下,当 Activity 再次被请求时,Activity 将重新被创建,原有的一些状态信息如滚动条位置、文本输入框中输入的内容将会丢失。

Activity 还有一个生命周期方法 onSaveInstanceState(BundleoutState),在系统对此 Activity 进行系统回收前调用,用来保存 Activity 的状态变量。此方法仅限于系统主动对此 Activity 进行回收,而不包括用户主动退出 Activity 的情况。因为在用户主动退出(例如单击返回按钮)的情况下,系统会默认当前的一些状态信息如编辑了一半的短信等是用户不需要的,主动放弃的。

6.1.2 Activity 的管理和状态管理

Android 通过栈的方式来管理 Activity,Activity 实例的状态决定它在栈中的位置。处于前台的 Activity 总是在栈的顶端,当前台的 Activity 因为异常或其他原因被销毁时,处于栈第二层的 Activity 将被激活,上浮到栈顶。当新的 Activity 启动入栈时,原 Activity 会被压入栈的第二层。Activity 在栈中的位置变化反映了它在不同状态间的转换。如图 6-2 所示。

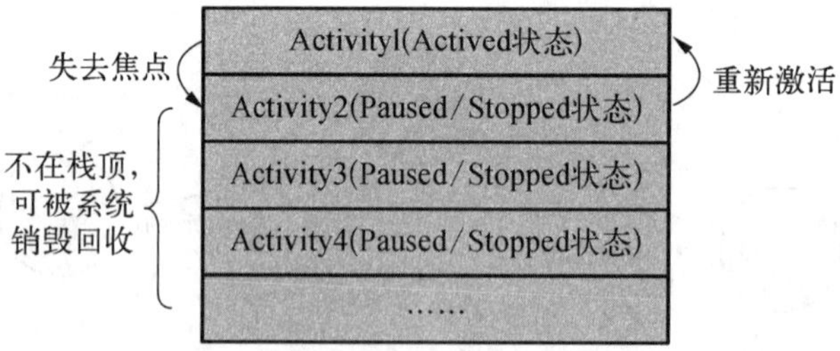

图 6-2 Activity 在栈中的状态转换

6.1.3 Activity 的运行

Activity 作为一个用户界面框架,它最大的功能就是提供一个窗口来显示用户界面。在 Activity 中可以通过添加运行的方法来控制 Activity 的运行,这些方法如下。

(1) onCreate():表示 Activity 正在被创建,进行初始化工作。例如调用 setContentView 加载资源、初始化 Activity 数据等。

(2) onRestart():表示 Activity 正在重新启动,从不可见的状态转变为可见的状态。

(3) onStart():表示 Activity 正在被启动,即将开始,在后台可见,前台不可见。

(4) onResume():表示 Activity 已经在前台显示出来并且可以操作了。

(5) onPause():表示 Activty 正在停止,一般情况下紧接着会调用 onStop。在特定情况下,如果这个时候快速回到当前的 Activity,那么 onResume 会被调用。

(6) onStop():表示 Activity 即将停止。

(7) onDestroy():表示 Activity 即将被销毁。

案例 6-1:Activity 生命周期演示

【微信扫码】案例 6-1 相关文件

新建一个 Android 项目,命名为 Intent_Teach_LifeCycle。

设计一个主 Activity 和两个子 Activity。通过主 Activity 启动不同的子 Activity,在过程中查看三个 Activity 的状态变化。

步骤 1:在 activity_main.xml 中编写布局文件,放置两个按钮,用来启动其他 Activity,代码如下。

```
<Button android: id = "@ + id /start_normal_activity"   android: text = "启动第一个界面" ······ />
<Button android: id = "@ + id /start_dialog_activity"   android: text = "启动第二个界面" ······ />
```

在 res/layout 中新增两个布局文件,分别命名为 normal_layout.xml 和 dialog_layout.xml,里面放置一个 TextView,设置 android:text="这是你看见的第一(二)个界面" />。

步骤 2:修改 MainActivity.java 主程序文件,添加 Activity 的生命周期的方法,运行时在 Log 中显示信息,代码如下。

```
public class MainActivity extends Activity {
    public static final String TAG = "MainActivity";
    private Button startNoramlActivity, startDialogActivity;
      @Override
      protected void onCreate(Bundle savedInstanceState) {
            ............
            Log.d(TAG, "MainActivity onCreate");
            startNoramlActivity = (Button) findViewById(R.id.start_normal_activity);
            startNoramlActivity.setOnClickListener(new OnClickListener() {
             @Override
             public void onClick(View v) {
                 Intent intent = new Intent(MainActivity.this, NormalActivity.class);
                 startActivity(intent);}  });
           ······ //startDialogActivity 实代码同 startNoramlActivity }
      @Override
      protected void onStart() {super.onStart();Log.d(TAG, "MainActivity onStart"); }
      ······ //按照 onStart 完成 onResume, onPause, onStop, onRestar, onDestroy 代码}
```

步骤 3:新建两个 Activity,分别为 NormalActivity 和 DialogActivity,代码和 MainActivity 相似,添加 Activity 的生命周期的方法,运行时在 Log 中显示信息,代码略。

运行程序,分别从 MainActivity 跳转到 NormalActivity,从 NormalActivity 返回

MainActivity，再跳转到 DialogActivity，从 DialogActivity 返回 MainActivity，最后退出程序。观察 Logcat 中的信息，如表 6-1 所示。

表 6-1　Activity 的生命周期变化

1. 首次启动 MainActivity	4. 从 MainActivity 启动界面 DialogActivity
02/17 10:14:39:Launching App D/MainActivity:MainActivity onCreate D/MainActivity:MainActivity onStart D/MainActivity:MainActivity onResume	D/MainActivity:MainActivity onPause D/DialogActivity:DialogActivity onCreate D/DialogActivity:DialogActivity onStart D/DialogActivity:DialogActivity onResume D/MainActivity:MainActivity onStop
2. 从 MainActivity 启动界面 NormalActivity	5. 从 DialogActivity 中的返回 MainActivity
D/MainActivity:MainActivity onPause D/NormalActivity:NormalActivity onCreate D/NormalActivity:NormalActivity onStart D/NormalActivity:NormalActivity onResume D/MainActivity:MainActivity onStop	D/DialogActivity:DialogActivity onPause D/MainActivity:MainActivity onReStart D/MainActivity:MainActivity onStart D/MainActivity:MainActivity onResume D/DialogActivity:DialogActivity onStop D/DialogActivity:DialogActivity onDestroy
3. 从 NormalActivity 中返回 MainActivity	6. 点击 MainActivity 中的退出按钮
D/NormalActivity:NormalActivity onPause D/MainActivity:MainActivity onReStart D/MainActivity:MainActivity onStart D/MainActivity:MainActivity onResume D/NormalActivity:NormalActivity onStop D/NormalActivity:NormalActivity onDestroy	D/MainActivity:MainActivity onPause D/MainActivity:MainActivity onStop D/MainActivity:MainActivity onDestroy

分析上述信息，总结 Activity 的生命周期的运行规律：① 首次启动：onCreate→onStart→onResume；② 再次打开：onRestart→onStart→onResume；③ 退出系统：onPause→onStop→onDestroy。

6.2　Intent

Intent 是一类特殊的组件，它负责对应用中的操作动作以及动作相关数据进行描述，Android 根据此描述，负责找到对应的组件，将 Intent 传递给此组件，并完成组件的调用。Intent 不仅可用于应用程序内部，也可用于应用程序之间的交互。

6.2.1　Intent 的作用

Intent 在 Android 中承担着指令传输的作用，实现 Activity 之间的交互与通信。

Intent 可以让应用程序向 Android 传递所需要的请求、数据等。而 Android 会根据 Intent 中所声明的内容自动选择所匹配的组件。由于 Intent 的出现，组件只要将自己的需求通过 Intent 进行描述，而不必具体实现对组件的引用，这些工作全部由底层的 Android Runtime 来实现，不仅在 Activity，在 Service、BroadcastReceiver 等都是通过 Intent 组件关联起来。Intent 最大的优点，就是完美地实现了调用者与被调用者之间的解耦。

6.2.2　Intent 的数据传递

Bundle 是以键值对的方式对数据进行保存的，类似于 Java 中的 Map。

Bundle 提供了一系列的 put、get 方法对数据进行保存和提取。其中 putAll(Bundle)方法是将传入的 Bundle 对象中的所有值存放于当前的 Bundle 中，它与 pubBundle(String key，Bundle value)是完全不同的，后者是将一个 Bundle 对象作为值进行保存，并不会取出其中的值进行保存。在构建好 Bundle 对象后，即可调用 Intent 对象的 putExtras 方法，将此 Bundle 对象放入 Intent 附加信息中，再传入新创建的目标组件。

1. 在 Intent 中传递简单数据

通过 data 属性：data 属性是一种 url，它可以指向我们熟悉的 http、ftp 等网络地址，也可以指向 ContentProvider 提供的资源。调用 Intent 的 setData 方法可以放入数据，调用 getData 方法可以取出数据。

通过 extra 属性：由于 data 属性只能传递数据的 url 地址，如果需要传递一些数据对象，就需要利用 extra 了。extra 可以通过 Intent 的 putExtra 方法放入数据，方法 putExtra 的参数是一个 Bundle 对象。

可以传递的数据分为八大基础类型，字符串和实现 Serializable 或 Parcelable 接口的对象，也可以传递 ArrayList 集合。

通过构建 Intent 对象，并以此为参数，调用 startActivity 方法，可以启动另外一个 Activity 组件。Android 还为 Activity 提供了另一个方法 startActivityForResult。不同于 startActivity，startActivityForResult 方法在启动新的 Activity 后，将在回调 onActivityResult 方法中获取新启动 Activity 的返回信息。

案例 6－2：多 Activity 中使用 Intent 传递数据

在本案例中将有两个界面，用来发送信息和接收信息。布局文件 send_main.xml 对应的 Activity 文件 Send.java 用来发送信息。布局文件 receive_main.xml 对应的 Activity 文件 Receive.java 用来接收和返回信息。新建一个 Android 项目，命名为 Intent_Teach_SendReceive。

【微信扫码】
案例 6－2
相关文件

步骤 1：在 res/layout 文件夹下新建发送信息的布局文件 send_main.xml，放置 Button 用来发送信息，TextView 用来接收反馈的信息，如图 6－3 所示，代码如下。

图 6－3　发送信息界面

```
<Button android:id="@+id/mybut"   android:text="@string/send_name" ……  />
<TextView android:id="@+id/msg"   android:text="接受第二个界面传回来的文字" ……  />
```

图 6-4　接收信息界面

步骤 2:在 res/layout 文件夹下新建接收信息的布局文件 receive_main.xml,放置 TextView 用来显示主程序发送的信息,两个 Button 用来返回不同类型的数据,如图6-4所示,代码如下。

```
<TextView android:id="@+id/show" ……  />
<Button android:id="@+id/retbut"   android:text="返回数据到 Send" ……  />
<Button android:id="@+id/retbut2"   android:text="返回其他数据到 Send" ……  />
```

步骤 3:在 src 下新建一个 Activity 文件,命名为 Send.java,作为发送信息程序,核心代码如下。

```
public class Send extends Activity {
    private Button mybut = null;private TextView msg = null;
    @Override
    protected void onCreate(Bundle savedInstanceState) {
        …………
        this.mybut.setOnClickListener(new OnClickListenerImpl()); }
    public class OnClickListenerImpl implements OnClickListener{
        @Override
        public void onClick(View v) {
            Intent intent = new Intent(Send.this, Receive.class);
            intent.putExtra("myinfo", "常熟理工学院移动开发课程实训");
            Send.this.startActivity(intent); //使用 stratActivity 处理没有返回信息的 Activity
            //如果有返回信息,使用 startActivityForResult(Intent intent,int requestCode)接受
            Send.this.startActivityForResult(intent,1); } }
  @Override
  protected void onActivityResult(int requestCode,int resultCode,Intent data){
    switch(resultCode){ //根据返回的结果显示不同的字符串
      case RESULT_OK:Send.this.msg.setText("确定的数据是" + data.getStringExtra
("retmsg"));break;
      case RESULT_CANCELED:Send.this.msg.setText("操作取消的数据是" + data.getStringExtra
("retmsg"));break;
        default:break; } }  }
```

步骤 4:在 src 文件夹下新建一个 Activity 文件,命名为 Receive.java,作为接收信息程序。

```
public class Receive extends Activity {
    private TextView show = null;private Button rebut = null;private Button rebut2 = null;
    @Override
    protected void onCreate(Bundle savedInstanceState) {
        ……
```

```
        Intent it = super.getIntent(); //使用 getIntent 方法获得前面传递过来的 Intent 数据
        String info = it.getStringExtra("myinfo"); //使用 String 获得 Intent 中为 myinfo 的数据
        //获得 show 、rebut 和 rebut2 组件,以及 rebut 和 rebut2 按钮的点击事件}
    public class OnClickListenerImpl implements OnClickListener{
      @Override
      public void onClick(View v) {
        Receive.this.getIntent().putExtra("retmsg", "确定!确定!我是由 RESULT_Ok 返回的数
据"); //回传 RESULT_OK 指示常量的数据
         //设置返回的数据和数据标识的常量
        Receive.this.setResult(RESULT_OK,Receive.this.getIntent());
        Receive.this.finish();} }
    public class OnClickListenerImpl2 implements OnClickListener{
      ...... //同 OnClickListenerImpl,RESULT_OK 换成 RESULT_CANCELED }}
```

运行程序,点击返回数据到 SEND,结果如图 6－5 所示,返回取消数据结果如图 6－6 所示。

发送INTENT的ACTIVITY程序
确定的数据是确定！确定！我是由RESULT_Ok返回的数据

图 6－5　返回确定数据的效果

发送INTENT的ACTIVITY程序
操作取消的数据是放弃！放弃！我是由RESULT_CANCELED返回的数据

图 6－6　返回取消数据的效果

2. 在 Intent 中传递复杂对象

在 Intent 中如果需要传递复杂对象,比如图片等,有两种方法:第一种方法使用 Bundle. putSerializable(Key,Object);第二种方法使用 Bundle. putParcelableArrayListExtra (Key, Object)。这些 Object 是不同的,前者使用 Serializable 接口,而后者使用 Parcelable 接口。推荐使用 Parcelable 接口,它不但可以利用 Intent 传递,还可在远程方法调用中使用。

实现 Parcelable 有两种方法:writeToParcel 方法,将类数据写入外部提供 Parcel 中,声明为 writeToParcel(Parcel dest,int flags);escribeContents 方法,返回描述信息的资源 ID,一般为零。

静态的 Parcelable.Creator 接口有两种方法:createFromParcel(Parcel in) 实现从 parcle 实例中创建出类的实例的功能;newArray(int size) 创建一个类型为 T,长度为 size 的数组。

案例 6－3:Intent 中传递复杂对象

在本案例中,设计一个注册的界面,注册信息包括名称、密码等字符串信息,还包括用于显示头像的图片数据。用户输入字符串信息和确定头像之后,点击注册,用户的信息将显示在另外一个界面上。新建一个 Android 项目,命名为 Intent_Teach_UserRegister。

【微信扫码】
案例 6－3
相关文件

步骤 1:在 res/values 中修改 strings.xml,编写头像名称资源文件,添加代码如下。

```
<string-array name="username">
        <item>校长</item><item>小红</item><item>小梁</item>…… //一共9个名字
</string-array>
```

步骤 2:在 res/layout 中修改 activity_main.xml,编写注册界面的布局文件,界面上左侧放置一组文本框和输入框,右侧放置 ImageView 用来显示头像,如图 6-7 所示,核心代码如下。

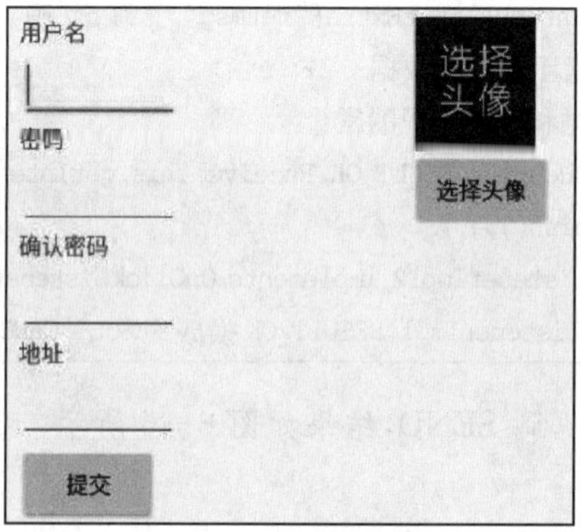

图 6-7　注册界面

```
<LinearLayout android:id="@+id/linearlayoutLeft" android:orientation="vertical" ……>
    <TextView android:id="@+id/textView1"  android:text="用户名" ……  />
    <EditText android:id="@+id/user" android:minWidth="200px" ……  />
    …… //添加三组 TextView 和 EditText 组合,分别用作密码、确认密码和地址
    <Button android:id="@+id/submit"  android:text="提交" ……  />
</LinearLayout>
<LinearLayout android:id="@+id/linearlayoutRight" android:orientation="vertical" ……>
  <ImageView android:id="@+id/imageView1" android:src="@drawable/blank" ……  />
  <Button android:id="@+id/button1"  android:text="选择头像" ……  />
</LinearLayout>
```

步骤 3:在 res/layout 中新增 head.xml 作为头像选择界面的布局文件,使用 GridView 控件,按照每行三个的排列方式放置头像图片,如图 6-8 所示,核心代码如下。

图 6-8　头像选择界面

```
< GridView android:id = "@ + id /gridView1" android:numColumns = "3" …… />
```

步骤 4：在 res/layout 中新增 register.xml 作为注册信息显示界面布局，核心代码如下。

```
< TextView android:id = "@ + id /user" android:text = "用户的名字" …… />
…… //添加三个 TextView，作为用户密码，确认密码和头像名字
< ImageView android:id = "@ + id /iconimage"  android:padding = "10dp" …… />
```

步骤 5：在 res 中新增 HeadActivity.java 编写头像选择程序，代码如下。

```
public class HeadActivity extends Activity {
   …… //获得资源文件中的头像资源
   @Override
   protected void onCreate(Bundle savedInstanceState) {
      …… //获得布局文件和控件
      BaseAdapter adapter = new BaseAdapter() { //使用 BaseAdapter 作为数据适配器
      @Override
       //使用 BaseAdapter 的 getView 方法实现适配器视图的创建和配置
      public View getView(int position, View convertView, ViewGroup parent){
            ImageView imageview; //因为是图像，所以使用 ImageView
            if  (convertView = = null){ //判断 View 是否存在，不存在的话需要生成和配置
               imageview = new ImageView(HeadActivity.this); //生成视图对象
               imageview.setAdjustViewBounds(true); //配置 View 的参数
               imageview.setMaxWidth(158);imageview.setMaxHeight(158);
               imageview.setPadding(5, 5, 5, 5);
            }else{ //如果存在视图对象，用 convertView 转换为具体的视图
            imageview = (ImageView)convertView; }
            imageview.setImageResource(imageId[position]); //图像加载到 ImageView 中
            return imageview; //返回加载了头像的 ImageView}
        …… //实现 BaseAdapter 其他三个抽象方法 getItemId() getItem() getCount() };
     gridview.setAdapter(adapter); //将 BaseAdapter 装配到控件上
     gridview.setOnItemClickListener(new OnItemClickListener(){ //设置头像的点击事件
      Override
      public void onItemClick(AdapterView<?> parent, View view, int position, long id){
        Intent intent = getIntent();Bundle bundle = new Bundle();
        bundle.putInt("imageId", imageId[position]); //将头像 Int 值装载到容器 Bundle 中
        bundle.putInt("imageIndex", position); //将头像 Index 值装载到容器 Bundle 中
        intent.putExtras(bundle); //将数据容器加载到 Intent 上
        setResult(0x11, intent); //设置前面的 Intent 接收的暗语 0x11
        finish();} }); } }
```

步骤 6：修改 MainActivity.java 编写主程序，效果如图 6－9 所示，核心代码如下。

```
public class MainActivity extends Activity {
    private int imageId = 0;private int imageindex = 0;
    @Override
    protected void onCreate(Bundle savedInstanceState) {
        …… //获得布局文件和控件
        button.setOnClickListener(new OnClickListener() { //点击之后跳转到头像界面
            @Override
            public void onClick(View v) { //当接收到 0x11 时接收返回的信息
              Intent intent = new Intent(MainActivity.this,HeadActivity.class);
              startActivityForResult(intent, 0x11);} });
        …… //获得 Submit 控件,注册其按钮点击事件 OnClickListenerImpl }
      public class  OnClickListenerImpl implements  OnClickListener{
        @Override
        public void onClick(View v) {
          String user = (findViewById(R.id.user)).getText().toString();
          …… //以此类推获得 pwd、repwd、email 三个输入内容
          if(!"".equals(user)&&!"".equals(pwd)&&!"".equals(repwd)){
              if (!pwd.equals(repwd)){Toast.makeText(MainActivity.this, "两个密码不一致",
Toast.LENGTH_LONG).show(); //检验信息合法性,包括输入不能为空、两个密码要一致等
              …… //清空 pwd 和 repwd 控件内容,并让 pwd 控件获得焦点
            }else{ Intent intent = new Intent(MainActivity.this, RegisterActivity.class);
            Bundle bundle = new Bundle(); //使用 Bundle 作为容器装载数据
            bundle.putCharSequence("user",user);
            //....以此类推完成 pwd, email 数据装配;
            bundle.putCharSequence("iconname",MainActivity.this.getResources().getStringArray
(R.array.username)[imageindex]);bundle.putInt("iconId",imageId); //头像信息
            intent.putExtras(bundle);startActivity(intent); }
        }else{Toast.makeText(MainActivity.this, "有内容没有填写完整", Toast.LENGTH_
LONG).show();}  } }
    @Override
    protected void onActivityResult(int requestCode,int resultCode,Intent data){
      super.onActivityResult(requestCode, resultCode, data);
      if (requestCode = = 0x11 && resultCode = = 0x11){
        Bundle bundle = data.getExtras();imageId = bundle.getInt("imageId");
        imageindex = bundle.getInt("imageIndex");
        ImageView iv = findViewById(R.id.imageView1);iv.setImageResource(imageId);}}}
```

步骤 7:在 src 中新建一个 Activity 文件,命名为 Register.java,作为信息显示程序,代码如下。

```
public class RegisterActivity extends Activity {
    @Override
    protected void onCreate(Bundle savedInstanceState) {
        …………
```

```
Intent intent = getIntent();Bundle bundle = intent.getExtras();
TextView user = (TextView) findViewById(R.id.user);
user.setText("您的用户名是:" + bundle.getString("user"));
…… //以此类推显示 pwd 和 email 和 iconname 内容
ImageView iconId = (ImageView) findViewById(R.id.iconimage);
//根据头像的资源 Int 值,直接提取头像显示在界面上
iconId.setImageResource(bundle.getInt("iconId"));}}
```

运行程序,首先在 MainActivity 中输入注册信息,并选择头像,如图 6-9 所示。点击提交之后,信息和头像显示在另外一个界面上,如图 6-10 所示。

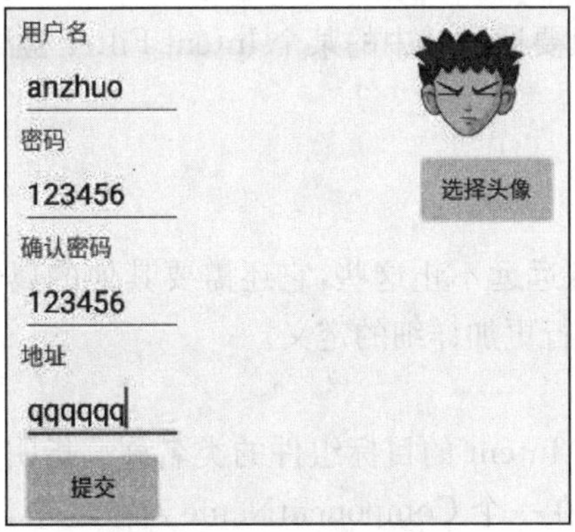

图 6-9　输入注册信息

图 6-10　注册信息显示界面

6.3　系统 Intent 组件和调用

6.3.1　系统 Intent 概述

系统 Intent 有显式和隐式之分。

显式 Intent:直接指定目标组件的类名,多用于应用程序内部组件之间的传递消息,比如 Service 或者 Activity。

隐式 Intent:未给出明确请求组件的名称,而是用 Intent 属性来描述所请求的意图。Android 则根据定义的 Intent Filter 来寻找最匹配的组件。这样的匹配是通过配置的 datatype、url、action 属性找到匹配的组件启动。

6.3.2　Intent Filter

Intent Filter 其实是一个 Intent 属性限制条件的集合,包含 Intent 对象的 action、data、category 等属性限制。Activity、Service、Broadcast Receiver 等组件为了告知 Android 能够处理哪些隐式 Intent,它们可以有一个或多个 Intent Filter。每个 Intent Filter 描述组件的一种能力,即能够接收的一组 Intent。

Intent Filter 筛掉不想要的 Intent,仅仅是不想要的隐式 Intent,因为显式 Intent 总是

能够传递到它的目标组件，不管它是否包含 Intent Filter。但对于隐式 Intent，仅当它能够通过组件的 Intent Filter 才能够传递给它。

Intent Filter 在 Android 应用程序的清单文件(AndroidManifest.xml)中以 intent-filter 节点的方式声明。但有一个例外，BroadcastReceiver 的 Intent Filter 通过调用 Context.registerReceiver()动态地注册，当以隐式方式调用 Intent 后，对其响应的组件是如何知道自己能响应、处理这个 Intent 的请求的呢？Android 系统中使用 IntentFilter 来定义组件所能响应的 Intent 请求类型、请求数据等。定义 IntentFilter 是在 AndroidManifest.xml 中使用 Intent-filter 标签来完成的，通过在 Intent-filter 节点中使用 action、category 等标签可以把相应的组件注册为 Intent 接收处理程序。Intent Filter 要检测隐式 Intent 的 action、data、category 这三个字段，其中任何一项失败，Android 系统都不会传递 Intent 给此组件。因为一个组件可以有多个 Intent Filter，Intent 只要通过其中的某个 Intent Filter 检测，就可以调用此组件。

6.3.3 Intent 重要属性和解析机制

Intent 在复杂应用中所需要提供的信息远远不止这些，它还需要其他的属性来对当前的 Intent 所描述的意图、要传递的数据等进行更加详细的定义。

构成 Intent 的属性有如下几点。

(1) ComponentName(组件名称)：指定 Intent 的目标组件的类名称。指明了将要处理的 Activity 程序，所有的组件信息都被封装在一个 ComponentName 对象之中，这些组件都必须在 AndroidManifest.xml 文件的"< Application >"中注册。

(2) Action(动作)：设置该 Intent 会触发的操作类型为字符串类型。可以通过 setAction()方法进行设置，在 Android 系统中已有一些表示 Action 操作的常量，如表 6－2 所示。

表 6－2　Action 操作的常量

Action 名称	描　述
ACTION_MAIN	作为一个程序的入口，不需要接收数据
ACTION_VIEW	用于数据的显示
ACTION_DIAL	调用电话拨号程序
ACTION_EDIT	用于编辑给定的数据
ACTION_PICK	从特定的一组数据之中进行数据的选择操作
ACTION_RUN	运行数据
ACTION_SENDTO	调用发送短信程序
ACTION_GET_CONTENT	根据指定的 Type 来选择打开操作内容的 Intent
ACTION_CHOOSER	创建文件操作选择器
ACTION_CALL	拨打电话

(3) Data(数据)：用来定义 Intent 中所需操作的数据。Android 系统中采用指向数据的 URI 表示数据。Android 系统中不同的应用使用的数据格式有不同的要求。要想 Android

系统能正确响应请求,正确的设置数据必不可少。对于不同的动作,其 URI 数据的类型是不同的。描述 Intent 所操作数据的 URI 及类型,可以通过 setData()进行设置,不同的操作对应着不同的 Data。

表 6 - 3　Data 格式表

操作类型	Data(URI)格式	范　例
浏览网页	http://网页地址	http://www.mldn.cn
拨打电话	tel:电话号码	tel:01051283346
发送短信	smsto:短信接收人号码	smsto: 13621384455
查找 SD 卡文件	file:///sdcard/文件或目录	file:///sdcard/mypic.jpg
显示地图	geo:坐标,坐标	geo:31.899533,-27.036173

(4) Type(数据类型):显式指定 Intent 的数据类型。一般 Intent 的数据类型能够根据数据本身进行判定,但是通过设置这个属性,可以强制采用显式指定的类型而不再进行推导,见表 6 - 4。

表 6 - 4　Type 指定的数据类型

作用	MIME 类型
发送短信	vnd.android-dir/mms-sms
设置图片	image/png
普通文本	text/plain
设置音乐	audio/mp3

(5) Category(类别):被执行动作的附加信息。用于指定执行的环境,描述执行操作的类别,通过 addCategory()方法设置多个类别,见表 6 - 5。

表 6 - 5　Category 的附加信息

Category 名称	描　述
CATEGORY_LAUNCHER	表示此程序显示在应用程序列表中
CATEGORY_HOME	显示主桌面,即开机时的第一个界面
CATEGORY_PREFERENCE	运行后将出现一个选择面板
CATEGORY_BROWSABLE	显示一张图片、E-mail 信息
CATEGORY_DEFAULT	设置一个操作的默认执行
CATEGORY_OPENABLE	Action 为"GET_CONTENT"时用于打开指定的 URI

(6) Extras(附加信息):包含所有附加信息的集合。每一个通过 startActivity 方法发出的隐式 Intent 都至少有一个 Category,就是 "android.intent.category.DEFAULT",所以所有希望接收隐式 Intent 的 Activity 都应该包括 "android.intent.category.

DEFAULT" 这一 Category 属性，不然将导致 Intent 匹配失败。传递一组键值对，可以使用 pubExtra()方法进行设置，主要功能是传递数据(URI)所需要的一些额外的操作信息，见表 6 - 6

表 6 - 6　Extras 附加信息表

操作数据	附加信息	作　用
短信操作	sms_body	表示要发送短信的内容
彩信操作	Intent.EXTRA_STREAM	设置发送彩信的内容
指定接收人	Intent.EXTRA_BCC	指定接收 E-mail 或信息的接收人
E-mail 收件人	Intent.EXTRA_EMAIL	用于指定 E-mail 的接收者，接收一个数组
E-mail 标题	Intent.EXTRA_SUBJECT	用于指定 E-mail 的标题
E-mail 内容	Intent.EXTRA_TEXT	用于设置 E-mail 内容

Intent 解析机制主要是通过查找已注册在 AndroidManifest.xml 中的所有 Intent Filter 及其中定义的 Intent 属性，最终找到匹配的 Intent。

Android 系统首先会根据调用方法的不同来确定定位何种类型的组件，是 Activity、Service 还是 BroadcastReceiver。这将大大缩小 Android 系统定位目标组件的范围。

(1) Action 检查：如果 Intent 指定了 Action，则目标组件的 Intent Filter 的 Action 列表中就必须包含有这个 Action，否则不能匹配；如果没有指定 Action，将自动通过检查。

(2) Category 检查：如果 Intent 指定了一个或多个 Category，将全部出现在列表中。

(3) Data 检测：每个 Data 元素指定数据类型(MIME Type)和一个 URI。URI 有四个属性：scheme、host、port、path，对应 URI 的格式为 scheme://host:port/path。

Data 检测既要检测 URI，也要检测 MIME Type，规则如下。

① 一个 Intent 对象既不包含 URI，也不包含 MIME Type：仅当 Intent Filter 不指定任何 URI 和 MIME Type 时，才不能通过检测，否则都能通过。

② 一个 Intent 对象包含 URI，但不包含 MIME Type：仅当 Intent Filter 不指定 MIME Type，同时它们的 URI 匹配，才能通过检测。例如，mailto:和 tel:都不指定实际数据。

③ 一个 Intent 对象包含 MIME Type，但不包含 URI：仅当 Intent Filter 只包含 MIEM Type 且与 Intent 相同，才通过检测。

④ 一个 Intent 对象既包含 URI，也包含 MIME Type：MIME Type 部分，与 Intent Filter 中之一匹配才算通过；URI 部分，Intent 对象的 URI 要出现在 Intent Filter 中或者它有 content:或 file: 或者 Intent Filter 没有指定 URI。

通过下面的案例学习如何使用 Intent 功能完成一些基本的操作。

案例 6 - 4：Intent 系统调用 Intent_Teach_IntentOperation

在本案例中将实现浏览器、拨号、发送短信、发送彩信、发送 E-mail 和显示联系人等功能。需要多个界面和程序文件，这些布局文件和程序对应关系如表 6 - 7 所示。

【微信扫码】
案例 6 - 4
相关文件

表 6-7　案例中布局文件和程序文件对应关系

布局文件(.xml)	程序文件(.java)	功能
activity_intentactionbrowse	IntentActionBrowseActivity	浏览器上网
layout_intentactiondial	IntentActionDial	调用拨号程序
activity_intentactionsendto	IntentActionSendToActivity	发送普通短信
activity_sendpicsms	IntentActionSendPicSmsActivity	发送彩信
activity_intentactionemail	IntentActionEmailActivity	发送 E-mail
/	IntentActionContactspeopleActivity	显示通信录中联系人

步骤 1:实现浏览网页功能,如图 6-11 所示,核心代码如下。

```
public class IntentActionBrowseActivity extends Activity {
    …………
    private class OnClickListenerImpl implements OnClickListener{
        @Override
        public void onClick(View v) {
            Uri uri = Uri.parse("http://www.cslg.cn") ; //设置操作的路径
            Intent it = new Intent();it.setAction(Intent.ACTION_VIEW); //设置操作的 Action
            it.setData(uri); //设置数据
            IntentActionBrowseActivity.this.startActivity(it) ; }   }  }
```

步骤 2:实现拨打电话功能,核心代码如下。

```
public class IntentActionDial extends Activity {
    …………
    private class OnClickListenerImpl implements OnClickListener{
        @Override
        public void onClick(View v) {
            String telStr = IntentActionDial.this.tel.getText().toString();
            Uri uri = Uri.parse("tel:" + telStr); //设置操作的路径
            Intent it = new Intent();it.setAction(Intent.ACTION_CALL); //设置操作的 Action
            it.setData(uri); //设置数据
            IntentActionDial.this.startActivity(it) ; }   } }
```

模拟器不能直接拨打真实电话,可以安装两个模拟器,在模拟器之间进行通信。一般第一个模拟器号码是 5554,第二个模拟器号码是 5556。模拟器拨打电话如图 6-12 所示。

图 6-11　网页浏览功能

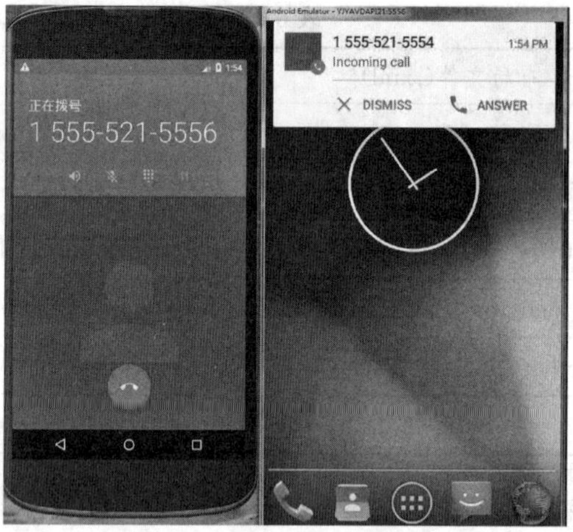

图 6-12　使用模拟器拨打电话

步骤 3:实现发送短信功能,核心代码如下。

```
public class IntentActionSendToActivity extends Activity {
   …………
   private class OnClickListenerImpl implements OnClickListener {
      @Override
      public void onClick(View v) {
        …………
         Uri uri = Uri.parse("smsto:" + telStr); //设置操作的路径
         Intent it = new Intent();it.setAction(Intent.ACTION_SENDTO); //设置操作的 Action
         it.putExtra("sms_body",note); //设置短信内容
         it.setType("vnd.android-dir/mms-sms"); //设置短信的 MIME 类型
         it.setData(uri); //设置数据
         IntentActionSendToActivity.this.startActivity(it); }}}
```

运行程序,模拟器 5554 发送短信到模拟器 5556,效果如图 6-13 所示。

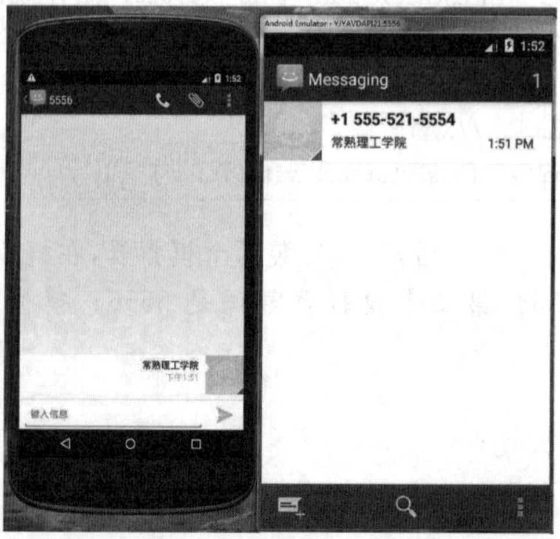

图 6-13　模拟器之间发送短信

步骤 4:实现发送彩信功能,核心代码如下。

```
public class IntentActionSendPicSmsActivity extends Activity {
   …………
   private class OnClickListenerImpl implements OnClickListener{
      @Override
      public void onClick(View v) {
        Uri uri = null;
        Intent it = new Intent();it.setAction(Intent.ACTION_SEND); //设置操作的 Action
        it.putExtra("address","5556"); //设置信息发送地址
        it.putExtra("sms_body","常熟理工学院计算机学院") ; //设置信息文本内容
        it.putExtra(Intent.EXTRA_STREAM,uri);it.setType("image/png"); //设置图片地址和
格式
        IntentActionSendPicSmsActivity.this.startActivity(it) ;     }  } }
```

运行程序,输入发送号码,发送内容之后,点击左上角的添加附件按钮,弹出选择框如图 6-14 所示。确定图片之后,单击发送,效果如图 6-15 所示。

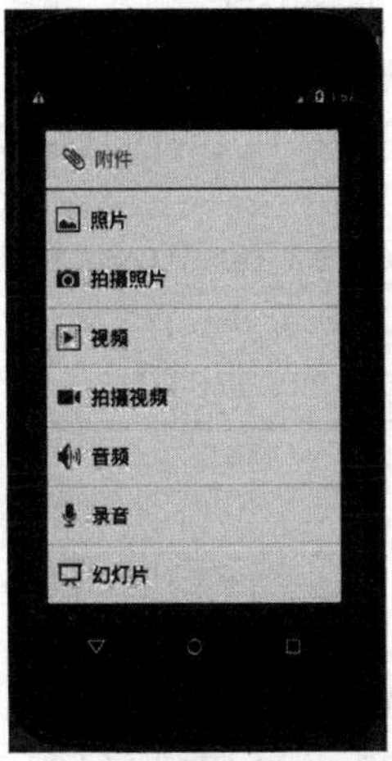

图 6-14　发送彩信前选择图片

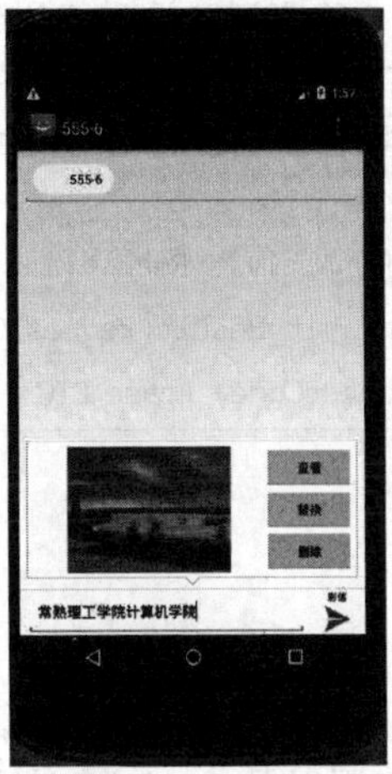

图 6-15　发送彩信

使用模拟器发送彩信,需要模拟器中预先有图片,可以按照如下操作:提前设置好 sdcard,启动模拟器。在 DDMS 界面中打开 mnt/sdcard 文件夹,点击窗口右上角的 PUSH 按钮,选择上传的图片。在 sdcard/DCIM/Camera/下的图片是系统摄像头拍摄的照片。上传的图片只需保存在 mnt/sdcard/下就可以了。在模拟器的主菜单中,选择 Dev Tools,然后选择 Media Scanner,系统就会刷新 sdcard 中的图片和媒体文件了。

步骤 5:实现发送 E-mail 功能,核心代码如下。

```
public class IntentActionEmailActivity extends Activity {
   …………
   private class OnClickListenerImpl implements OnClickListener{
      @Override
```

```
        public void onClick(View v) {
            Intent emailIntent = new Intent(Intent.ACTION_SEND) ;
            emailIntent.setType("plain/text") ; //设置类型
            String address[] = new String[] {"42598585@qq.com"} ; //设置邮件地址
            String subject = "邮件标题";String content = "邮件内容!" ;
            emailIntent.putExtra(Intent.EXTRA_EMAIL, address);
            emailIntent.putExtra(Intent.EXTRA_SUBJECT, subject);
            emailIntent.putExtra(Intent.EXTRA_TEXT, content);
            IntentActionEmailActivity.this.startActivity(emailIntent);}}}
```

步骤 6:实现显示联系人功能,核心代码如下。

```
public class IntentActionContactspeopleActivity extends Activity {
    private static final int PICK_CONTACT_SUBACTIVITY = 1; //设置标志信息
    @Override
    public void onCreate(Bundle savedInstanceState) {
        ............
        Uri uri = Uri.parse("content://contacts/people") ;
        Intent intent = new Intent(Intent.ACTION_PICK, uri);
        super.startActivityForResult(intent, PICK_CONTACT_SUBACTIVITY) ;    }
    @Override
   protected void onActivityResult(int requestCode, int resultCode, Intent data) {
        super.onActivityResult(requestCode, resultCode, data);
        switch(requestCode) {case PICK_CONTACT_SUBACTIVITY: } }}
```

小结

本章主要讲解了 Activity 的相关知识,包括 Activity 入门、Activity 生命周期、Activity 启动模式、Intent 的使用以及 Activity 中的数据传递,在应用程序中凡是有界面都会使用到 Activity,因此,要求初学者必须熟练掌握该组件的使用。

课后练习

【微信扫码】
第 6 章课后练习

第 7 章

广播处理、数据共享和权限管理

BroadcastReceiver 是 Android 系统的四大组件之一，可以方便地实现系统中不同组件之间的通信。它采用异步传输信息的机制，相当于收音机，而许许多多的广播电台就是消息的发送者，它通过特定的频率来发送。

7.1 广播机制简介

Android 系统中的广播机制非常灵活。Android 系统中的每个应用程序都可以对自己感兴趣的广播进行注册，这样该程序就只会接收到自己所关心的广播内容，这些广播可能是来自系统的，也可能是来自其他应用程序的。Android 系统允许应用程序自由地发送和接收广播。

广播消息在本质上就是一个 Intent 对象。在 Android 系统中，发送和接收的工作分别由 sendBroadcast 方法和注册的 BroadcastReceiver 类完成。只有当发送和接收都具有相同频率时，接收端才能接收到发送的信息，而这里的频率就是 Intent 对象中的 Action 属性。接收广播的方法则需要广播接收器。使用广播接收器需要继承类 BroadcastRecevier，并实现里面的 onReceive()方法。广播接收器是不允许使用线程的，因此不要在里面添加复杂的逻辑操作。

为了让广播接收器能够接受广播，还需要对感兴趣的广播进行注册。注册广播的方法有两种：一是静态注册，二是动态注册。如果需要程序在未启动的情况下接收消息，就必须使用静态注册方法。使用静态广播接收必须要在 AndroidManifest.xml 文件中注册才可以使用。注册的广播会在 Application 标签内出现一个新的标签 receiver，所有的静态广播接收器都必须在这里进行注册。同时还需要在 receiver 下面的 intent-filter 标签内添加对应的 action。

静态注册虽然使用方便，但是当程序运行中需要灵活地接收广播消息，就必须让广播接收器对接收的消息使用动态注册方法。广播接收器的动态注册需要继承类 BroadcastRecevier，并重写父类的 onReceive()方法。在 onReceive()方法，首先创建一个 IntentFilter，添加对应的 action，这个就是接收器想要监听的广播。然后创建一个 NetworkChangeReceiver 的实例，调用 registerReceiver()方法进行注册，将 IntentFilter 实例和 NetworkChangeReceiver 实例都传递进去，这样就能接收到对应的广播消息了。停止接收动态注册的广播消息必须取消注册才行，取消注册在 onDestroy()方法中通过 unregisterReceiver()方法来实现。

7.1.1 系统广播

Android 内部有很多系统级别的广播，比如电池的电量变化状况等，可以在应用程序中通过监听这类广播消息来获得系统的状态信息。下面通过一个案例来介绍系统广播的使用。

案例 7－1:使用 BroadcastReceiver 实现系统监听网络状态变化

【微信扫码】
案例 7－1
相关文件

新建一个 Android 项目,命名为 BR_Teach_SysBroadcast Receiver。当手机所连接的网络发生变化时,通过 BroadcastReceiver 接收广播,并显示提示信息。

步骤 1:创建继承 BroadcastReceiver 的类 MyReceiver,实现 onReceive()方法。

```
public class MyReceiver extends BroadcastReceiver {
    @Override
    public void onReceive(Context context, Intent intent){ …… }}
```

步骤 2:在 AndroidManifest.xml 中,首先在< manifest ></manifest >添加系统广播和访问网络两个权限,代码如下。

```
< uses-permission android:name = "android. permission. RECEIVE_BOOT_COMPLETED" />
< uses-permission android:name = "android. permission. ACCESS_NETWORK_STATE" />
```

在< Application ></ Application >中静态注册广播接收者。

```
 < receiver android:name = ". MyReceiver"> //指定具体注册广播接收者为 MyReceiver
     < intent - filter android:priority = "1000"> //广播的优先级别为 1000
         < action android: name = " android. intent. action. BOOT _ COMPLETED"  /> //确定接收
Action 名为 android. intent. action. BOOT_COMPLETED 的广播
     < /intent-filter >< /receiver >
```

步骤 3:修改 MainActivity 实现监听网络变化。

```
public class MainActivity extends AppCompatActivity {
    private IntentFilter intentFilter;private NetworkChangeReceiver networkChangeReceiver;
    @Override
  protected void onCreate(Bundle savedInstanceState) {
     …………
     intentFilter = new IntentFilter(); //创建 Intent Filter,设置监听的 Action
     intentFilter. addAction("android. net. conn. CONNECTIVITY_CHANGE");
     networkChangeReceiver = new NetworkChangeReceiver(); //注册监听事件
     registerReceiver(networkChangeReceiver, intentFilter);}
    @Override
    protected void onDestroy() { //程序退出时,注销监听
       super. onDestroy();unregisterReceiver(networkChangeReceiver);}
    //创建一个继承于 BroadcastReceiver 的 NetworkChangeReceiver
    class NetworkChangeReceiver extends BroadcastReceiver {
        @Override
        public void onReceive(Context context, Intent intent) {
           Toast. makeText(context, "网络状态发生变化" + getNetworkType(), Toast. LENGTH_
LONG). show();} }
```

```
public String getNetworkType(){ //获得当前网络的状态
        String netTypeString = null;
        ConnectivityManager connectivityManager = (ConnectivityManager) getSystemService
(Context.CONNECTIVITY_SERVICE); //通过 getSystemService 获得 NetworkInfo 实例
        NetworkInfo  info = connectivityManager.getActiveNetworkInfo();
        if (info = = null) {netTypeString = "没有网络连接";return netTypeString;}
        int nType = info.getType();
        if (nType = = ConnectivityManager.TYPE_MOBILE) {
            String extraInfo = info.getExtraInfo();
            if (!TextUtils.isEmpty(extraInfo)) {
                if (extraInfo.toLowerCase().equals("cmnet")) {netTypeString = "连接 CMNET 网
络";}else{netTypeString = "连接 CMWAP 网络";}}
        }else if(nType = = ConnectivityManager.TYPE_WIFI) {netTypeString = "连接 WIFI";}
        return netTypeString;}}
```

运行程序，点击飞行模式时显示如图 7-1 所示，取消飞行模式自动连接网络显示如图 7-2 所示，点击 Wi-Fi 显示如图 7-3 所示。

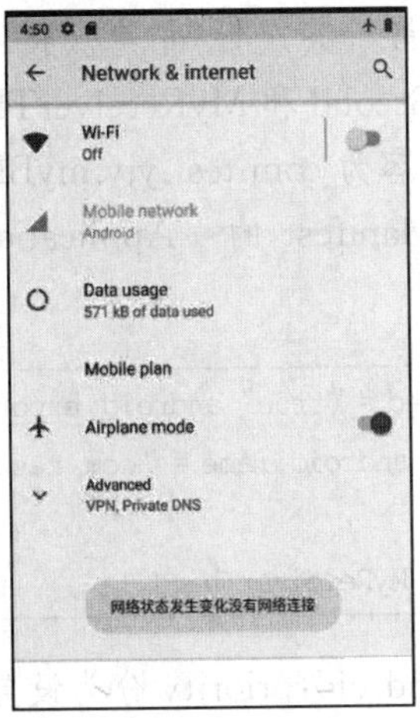

图 7-1 处于飞行模式

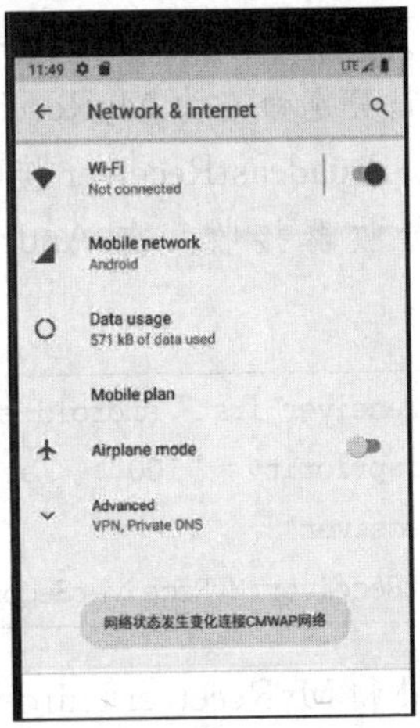

图 7-2 处于移动网络模式

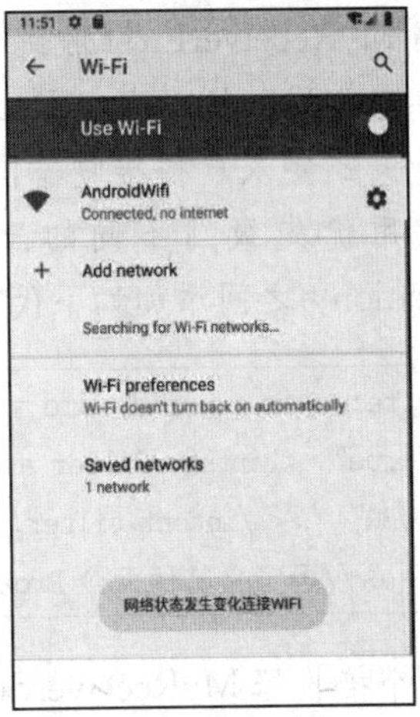

图 7-3 处于 Wi-Fi 模式

7.1.2 自定义广播

除了系统广播，用户如何在应用程序中发送自己定义的广播呢？这类的广播可以分为两种类型：普通广播和有序广播。这两种广播的区别如下。

普通广播(Normal Broadcast)：是一种完全异步执行的广播。当一条广播消息发出之后，所有的广播接收器都会在同一时间接收到这条消息。广播没有先后顺序，消息传递的效率比较高，无法被截取。缺点是接收者不能将处理结果传递给下一个接收者，并且无法终止传播。Context 提供 sendBroadcast()方法用于发送普通广播。

有序广播(Ordered Broadcast)：是一种同步执行的广播。广播接收有先后顺序。当某一条消息发出之后，在某一时刻，只有一个广播接收器收到这条消息，当这个广播接收器处理完这条消息之后，广播才会继续传递。接收者将按预先声明的优先级依次接收。如：A 的级别高于 B，B 的级别高于 C，那么，Broadcast 先传给 A，再传给 B，最后传给 C。优先级别声明在 intent-filter 元素的 Android：priority 属性中，数越大优先级别越高，取值范围为 −1 000～1 000。优先级别也可以调用 IntentFilter 对象的 setPriority（）进行设置。Ordered Broadcast 接收者可以终止 Broadcast Intent 的传播，传播一旦终止，后面的接收者就无法接收到 Broadcast。这样的广播缺点在于广播容易被接收优先级别高的接收器截取，优先级别低的接收器将收不到广播消息。Context 提供 sendOrderedBroadcast()方法用于发送有序广播。下面通过案例来介绍自定义广播的使用。

【微信扫码】
案例 7－2
相关文件

案例 7－2：自定义广播的使用

新建一个 Android 项目，命名为 BR_Teach_SelfBroadcastReceiver。

步骤 1：新建 BroadcastReceiver，命名为 MyReceiverFirst。

```
public class MyReceiverFirst extends BroadcastReceiver {
  @Override
  public void onReceive(Context context, Intent intent) { ……}
```

当 MyReceiverFirst 收到自定义广播，就会弹出“收到的第一条广播”信息。按照这个方法再定义两个 BroadcastReceiver，分别命名为 MyReceiverSecond 和 MyReceiverThird。

步骤 2：定义广播接收器信息。BroadcastReceiver 接收内容为 com.test.yjy.myBR 的广播。使用优先级别不同的三个接收器接收。在 AndroidManifest 的< Application ></Application >之间添加如下代码。

```
< receiver android: name = ". MyReceiverFirst" android: enabled = "true" android: exported =
"true">< intent-filter android: priority = "100">< action android: name = "com. test. yjy.
myBR" />< /intent-filter >< /receiver >
…… //再定义其他两个 BroadcastReceiver:MyReceiverSecond 和 MyReceiverThird
```

两个接收器 MyReceiverSecond 和 MyReceiverThird 的 android：priority 分为设置为 50 和 0，同时添加权限。

```
< uses-permission android:name = "android. permission. RECEIVE_BOOT_COMPLETED" />
< uses-permission android:name = "android. permission. ACCESS_NETWORK_STATE" />
```

步骤 3：修改布局文件，添加两个按钮，用来启动有序和无序广播。

```
< Button android:onClick = "clickOrderBroadcast"  …… />
< Button android:onClick = "clickBroadast"  …… />
```

步骤 4：修改 MainActivity 代码，主要实现钮点击事件中加入发送自定义广播的逻辑。

```
public class MainActivity extends AppCompatActivity {
    ............
    public void clickBroadast(View view) {
        Log.i("MainActivity", "无序广播");
        //构造一个 Intent,将要发送的 com.test.yjy.myBR 广播值加入 Intent 中
        Intent intentunorder = new Intent("com.test.yjy.myBR");
        sendBroadcast(intentunorder);} //调用 sendOrderedBroadcast()发送有序广播
    public void clickOrderBroadcast(View view) {
        Log.i("MainActivity", "有序广播");
        Intent intentorder = new Intent("com.test.yjy.myBR");
        sendOrderedBroadcast(intentorder,null);}} //调用 sendBroadcast()发送无序广播
```

7.1.3 本地广播

为了解决广播的安全性问题，Android 系统引入了一套本地广播机制，通过这套机制，发出的广播只能在应用程序的内部进行传递，并且广播接收器也只接收来自本地应用程序发出的广播，这样广播的安全性就得到了保障。

本地广播消息使用 LocalBroadcastManager 进行管理。本地广播需要在程序启动之后才能使用，所以只能动态注册，而不能静态注册。

【微信扫码】案例 7－3 相关文件

案例 7－3：本地广播的使用

新建一个 Android 项目，命名为 BR_Teach_LocalBroadcastReceiver。

步骤 1：在 AndroidManifest.xml 添加如下的权限。

```
< uses-permission android:name = "android.permission.ACCESS_NETWORK_STATE" />
< uses-permission android:name = "android.permission.RECEIVE_BOOT_COMPLETED" />
```

步骤 2：新建一个继承 BroadcastReceiver 的新类 MyLocalReceiver，用来接收广播。

```
public class MyLocalReceiver extends BroadcastReceiver {
    @Override
    public void onReceive(Context context, Intent intent) { ……}}
```

步骤 3：修改 MainActivity 代码，实现本地广播功能。

```
public class MainActivity extends AppCompatActivity {
  …… (声明 IntentFilter、MyLocalReceiver、LocalBroadcastManager 三个对象)
  @Override
  protected void onCreate(Bundle savedInstanceState) {
     …… //实例化 LocalBroadcastManager,通过 getInstance()方法获得实例
   localBroadcastManager = LocalBroadcastManager.getInstance(this);
   ............
   button.setOnClickListener(new View.OnClickListener() {
```

```
    @Override
    public void onClick(View v) {
      Intent intent = new Intent("com.broadcasttest.LOCAL_BROADCAST");
      localBroadcastManager.sendBroadcast(intent); //通过 sendBroadcast 发送广播}});
      intentFilter = new IntentFilter();
      intentFilter.addAction("com.broadcasttest.LOCAL_BROADCAST");
      myLocalReceiver = new MyLocalReceiver();
       //registerReceiver()注册广播的接收器
      localBroadcastManager.registerReceiver(myLocalReceiver, intentFilter);}
  @Override
  protected void onDestroy() { //销毁
    super.onDestroy();localBroadcastManager.unregisterReceiver(myLocalReceiver);}}
```

运行程序,点击发送广播按钮,效果如图 7-4 所示。

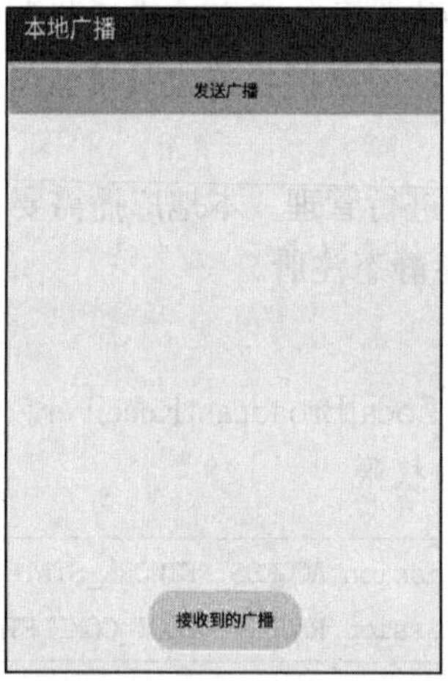

图 7-4　本地广播效果

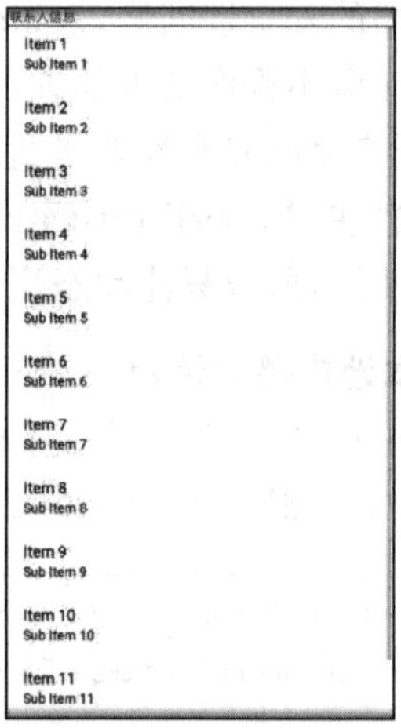

图 7-5　布局文件效果

7.2　Content Provider

Content Provider(内容提供器)的用法一般有两种,一种是使用现有的内容提供器来读取和操作相应程序中的数据,另一种是创建自己的内容提供器,作为外部访问接口,提供给程序的数据使用。

如果一个应用程序通过内容提供器对其数据提供了外部访问接口,那么任何其他的应用程序就都可以对这部分数据进行访问。Android 系统中自带的电话簿、短信、媒体库等程序都提供了类似的访问接口,这就使得第三方应用程序可以充分地利用这部分数据来实现更好的功能。下面我们就通过使用 Content Provider 来获取系统电话簿的功能。

【微信扫码】
案例 7-4
相关文件

案例 7-4:使用 Content Provider 获得通信录联系人信息

新建一个 Android 项目,命名为 DS_Teach_ContentProvider,因为读取联

系人属于系统权限，首先在 AndroidManifest.xml 中添加权限。

```
< uses-permission android:name = "android.permission.WRITE_CONTACTS"  />
< uses-permission android:name = "android.permission.READ_CONTACTS"  />
```

步骤 1：程序需两个布局，首先修改 activity_main.xml 作为主布局文件，主要放置一个 ListView，用来显示联系人代码如下。

```
< TextView android:text = "联系人信息"……  />
< ListView android:id = "@ + id /lv" android:paddingLeft = "5sp"…… />
```

联系人的信息一般包含姓名和电话号码，在 ListView 中还需要一个小布局结构显示。在 res/layout 中新建一个布局文件，命名为 list_item.xml，其中放置一个 TextView。

```
< TextView android:id = "@ + id /tv" ……  />
```

步骤 2：修改 MainActivity 代码。

```
public class MainActivity extends Activity {
   private static final String [ ] COLUMNS = { ContactsContract. Contacts. DISPLAY_ NAME,
ContactsContract. CommonDataKinds. Phone. NUMBER };
   private static final int PERMISSIONS_REQUEST_CODE = 0x00099;
     @Override
   protected void onCreate(Bundle savedInstanceState) {
        ……writeContact();}
   protected void writeContact(){ //获得系统权限
       if (Build. VERSION. SDK_INT> = Build. VERSION_CODES. M && checkSelfPermission(Manifest.
permission. READ_CONTACTS)! = PackageManager. PERMISSION_GRANTED){requestPermissions(new String
[]{Manifest. permission. READ_CONTACTS},PERMISSIONS_REQUEST_CODE); }else{
      Cursor cursor = this. managedQuery(ContactsContract. CommonDataKinds. Phone. CONTENT_
URI, COLUMNS, null, null, null);
   int idIndex = cursor. getColumnIndex(COLUMNS[0]);
   int numIndex = cursor. getColumnIndex(COLUMNS[1]);
   ArrayList<String> list = new ArrayList<String>();
   while(cursor. moveToNext()){
     String name = cursor. getString(idIndex);String num = cursor. getString(numIndex);
     list. add(name + " " + num);}
     if(cursor. isClosed()){ cursor. close();}
   ListView lv =  findViewById(R. id. lv);
   ArrayAdapter<String> adapter = new ArrayAdapter<String>(this, R. layout. list_item,
list);
     lv. setAdapter(adapter); }     }   }
```

7.3 Android 的权限管理机制

作为一个开放的系统，Android 的系统安全性一直是用户比较关心的。在之前的版本中，如果一个 App 在运行时需要一些系统权限，那么在安装时就需要被授予这些访问权限，如果用户不允许则 App 也不能被安装，这带来了极大的安全隐患。App 可以在用户不知晓的情况下访问权限内的各种信息。一些不法分子可以利用这样的缺陷，恶意获取系统权限并进行违法活动。

在 Android 6.0 之后的版本，Android 的安全体系被重新设计，采用了一种"运行时授权"的方案。用户在 App 安装的时候，不需要一次性授予全部的权限，而是可以在使用的时候，对某些权限申请进行授权。比如，一个 App 如果在运行时要求申请打开摄像头，就算拒绝了这个申请，也能使用这个 App 的其他功能，而不是直接无法安装。

7.3.1 权限的申请和程序处理方法

"运行时授权"运行的版本，需要在 targetSDKVersion≥23 就是 Android 6.0 以上。如果小于这个版本，Android 系统默认采用低版本的权限规则，如果是等于或者高于这个版本，Android 才会采用新的权限管理权限。

Android 将权限分为两种：第一种是危害不大的权限，比如访问网络的 Wi-Fi 状态等。和之前的请求方式一样，清单文件中注册之后，安装时就获得权限。表 7－1 大致列出了这些权限。第二种权限，就是涉及用户隐私或者设备安全性的权限，比如获得用户的设备联系人信息、定位设备的地理位置等，对于这部分权限，在运行前必须由用户手动点击授权即可，否则程序无法获得相应的功能。怎么知道哪些权限属于危险权限呢？可以访问 https://developer.android.google.cn/guide/topics/security/permissions。

表 7－1 普通权限表

权限组	权　限
CALENDAR	READ_CALENDAR | WRITE_CALENDAR
CAMERA	CAMERA
CONTACTS	READ_CONTACTS | WRITE_CONTACTS | GET_ACCOUNTS
LOCATION	ACCESS_FINE_LOCATION ACCESS_COARSE_LOCATION
MICROPHONE	RECORD_AUDIO
PHONE	READ_ PHONE _ STATE | CALL _ PHONE | READ _ CALL _ LOG | USE _ SIPWRITE_CALL_LOG |ADD_VOICEMAIL | PROCESS_OUTGOING_CALLS
SENSORS	BODY_SENSORS
SMS	SEND_SMS|RECEIVE_SMS|READ_SMS|RECEIVE_WAP_PUSH | RECEIVE_MMS
STORAGE	READ_EXTERNAL_STORAGE |WRITE_EXTERNAL_STORAGE

被定义为危险的权限数量比较多，如果在系统运行中很多权限需要授予，一个一个点击

授权是非常不方便的。为此，Android 系统在授权的设计上，当同组的任何一个权限被授予，其他的权限也会被自动授予。比如，当 READ_CONTACTS 被授予时，同 CONTACTS 组的 WRITE_CONTACTS 和 GET_ACCOUNTS 两个权限也会被授予。

程序中的权限授予分为两步：首先需要在 AndroidManifest. xml 中的 android：permission 属性中添加权限声明，然后在程序中请求权限和检查权限。在具体的实现中，有三种方法实现，分别是：直接编写代码进行授权；对权限申请进行轻量级封装；结合 RxJava 的权限库 RxPermissions 实现。

日常生活中，我们经常用手机来拍摄照片，这需要打开手机的摄像头。打开摄像头的权限名称为 CAMERA，因为打开摄像头会获得用户的隐私，因而被列为危险权限。我们就通过打开摄像头这个权限来学习运行时权限的使用方法。

【微信扫码】
案例 7-5
相关文件

案例 7-5：权限的多种处理方式案例

新建一个 Android 项目，命名为 PM_Teach_PermissionDemo。

步骤 1：首先打开 AndroidManifest.xml 文件，声明如下权限。

```
< uses-permission android:name = "android.permission.CAMERA" />
```

步骤 2：修改 activity_main.xml 文件，放置三个按钮用于不同的权限设置，如图 7-6 所示。

```
< TextView android:l android:text = "Android Studio 权限体系!" …… />
< Button android:id = "@ + id /bntpm" android:text = "直接进行赋权" …… />
< Button android:id = "@ + id /bntpmactivity" android:text = "使用 MPermissionsActivity"…… />
< Button android:id = "@ + id /bntRx" android:text = "使用 RxPermission" …… />
```

步骤 3：新建一个名为 activity_main_direct.xml 的布局文件，放置一个文本框用于显示授权类型和两个按钮用于启动摄像头和查看权限，如图 7-7 所示。

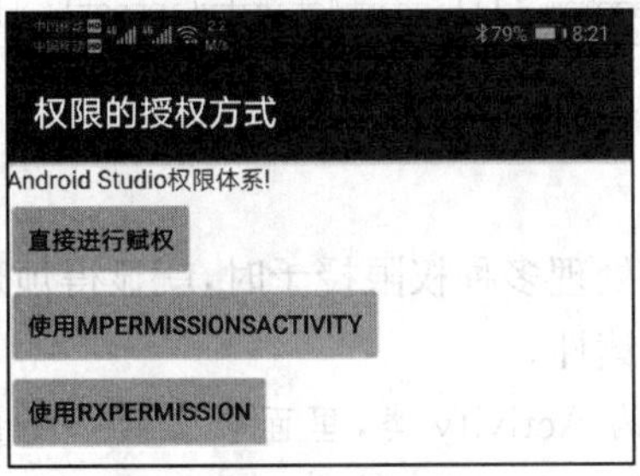

图 7-6　权限主界面

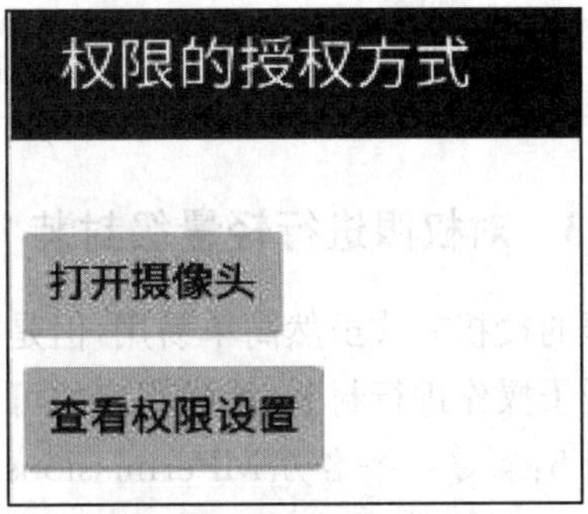

图 7-7　权限的授权方式

```
< TextView android:id = "@ + id /showtxt" android:textSize = "20sp" …… />
< Button android:id = "@ + id /bntcamera" android:text = "打开摄像头" …… />
< Button android:id = "@ + id /bntsetting" android:text = "查看权限设置" …… />
```

7.3.2 直接授权的实现

步骤 4:新建一个 Activity 类,命名为 MainActivity_Direct,直接进行授权修改,代码如下。

```
public class MainActivity_Direct extends AppCompatActivity {
  private Button bntcamera, bntsetting; private TextView showtxt;
  @Override
  protected void onCreate(Bundle savedInstanceState) {
    …………
    bntcamera.setOnClickListener(new View.OnClickListener() {
    @Override
    public void onClick(View v) { //使用 checkSelfPermission()判断是否授权
    if (ContextCompat.checkSelfPermission(MainActivity_Direct.this, Manifest.permission.
CAMERA)! = PackageManager.PERMISSION_GRANTED){ //如果没有授权则授权
      ActivityCompat.requestPermissions(MainActivity_Direct.this, new String[]{android.
Manifest.permission.CAMERA},1);}else{openCamera(); //如果已授权,进行相应操作}});
      private void openCamera() {…… //打开摄像头 }
      @Override
    public void onRequestPermissionsResult(int requestCode, String[] permissions, int[]
grantResults) { //授权函数的回调函数
     switch (requestCode) {
        case 1:if (grantResults.length > 0 && grantResults[0] = = PackageManager.PERMISSION
_GRANTED) { openCamera();} else { Toast.makeText(this, "你申请的权限被拒绝", Toast.LENGTH
_SHORT).show();} break;default: }}
   bntsetting.setOnClickListener(new View.OnClickListener() { //查看权限
   @Override
   public void onClick(View view) {
    Intent intent = new Intent(Settings.ACTION_APPLICATION_DETAILS_SETTINGS);
    intent.setData(Uri.parse("package:" + getPackageName()));startActivity(intent); }});}}
```

7.3.3 对权限进行轻量级封装

上面的授权方式虽然简单易用,但是如果需要处理多种权限授予时,就显得烦琐。可以对权限授予操作进行封装,增强程序的可靠性和可读性。

步骤 5:新建一个名为 MPermissionsActivity 的 Activity 类,里面包含两个变量参数和八个方法,下面列出变量和方法的名字,具体实现代码请参照对应的二维码内容。

```
public class MPermissionsActivity extends AppCompatActivity {
    private final String TAG = "MPermissions";
    private int REQUEST_CODE_PERMISSION = 0x00099;
    //检测权限是否都已授予
    public void requestPermission(String[] permissions, int requestCode){ ...... }
```

```
    //获取需要集中申请权限的列表
    private List<String> getDeniedPermissions(String[] permissions) { …… }
   public void onRequestPermissionsResult(int requestCode, @NonNull String[] permissions,
@NonNull int[] grantResults) { …… } //授权的回调函数@Override
  private boolean verifyPermissions(int[] grantResults){ …… } //验证是否获得授权
  private void showTipsDialog() { …… } //显示提示对话框
  private void startAppSettings() { …… } //启动当前应用设置页面
  public void permissionSuccess(int requestCode) { …… } //获得权限成功显示
  public void permissionFail(int requestCode) { …… }} //获得权限失败显示
```

步骤 6：新建一个 Activity 类，命名为 MainActivity_MPActivity，修改代码如下。

```
public class MainActivity_MPActivity extends MPermissionsActivity {
  private Button bntcamera,bntsetting;private TextView showtxt;private int mPInt;
  @Override
  protected void onCreate(Bundle savedInstanceState) {
    …………
    bntcamera.setOnClickListener(new View.OnClickListener() {
    @Override
    public void onClick(View v){
     //调用 MPermissionsActivity 中的 requestPermission()授权
    requestPermission(new String[]{Manifest.permission.CAMERA},0x001);}  });
    bntsetting.setOnClickListener(new View.OnClickListener() { ……打开权限设置页面 }
    public void permissionSuccess(int requestCode){ //授权成功
       super.permissionSuccess(requestCode);if(requestCode = = 0x001){openCamera();} }
    private void openCamera() { //打开摄像头
      try {Intent intent = new Intent(MediaStore.ACTION_IMAGE_CAPTURE);
      startActivity(intent);} catch (SecurityException e) {e.printStackTrace();}}}
```

7.3.4　结合 RxJava 的授权权限

上面的两种授权方式虽然功能齐全，但是看起来有点繁琐。如果想简化，可以引入第三方库 RxPermissions，它是参考 RxJava 设计，功能强大。下面我们使用它来授权。

步骤 7：在 app/build.gradle 的 dependencies 中添加依赖。

```
implementation 'io.reactivex:rxjava:1.1.6'
implementation 'com.tbruyelle.rxpermissions:rxpermissions:0.7.0@aar'
```

步骤 8：新建一个命名为 MainActivityRx 的 Activity 类，使用 RxPermissions 进行授权，核心修改代码如下。

```
public class MainActivityRx extends AppCompatActivity {
  private Button bntcamera,bntsetting;private TextView showtxt;
  @Override
  protected void onCreate(Bundle savedInstanceState) {
    …………
    bntcamera.setOnClickListener(new View.OnClickListener() {
      @Override
      public void onClick(View view) {
        RxPermissions.getInstance(MainActivityRx.this) //使用 RxPermissions 进行授权
          .request(Manifest.permission.CAMERA)
          .subscribe(new Action1<Boolean>() {
          @Override
        public void call(Boolean granted){if (granted){openCamera();}}}); }});
   bntsetting.setOnClickListener(new View.OnClickListener() {…… //打开权限设置页面}
   private void openCamera() {…… //打开摄像头}   }
```

可以看到,使用第三方库进行授权在实现上非常简单。

小结

本章内容分为三个部分。第一部分讲解广播接收器的相关知识,包括什么是广播接收器,自定义广播以及广播的类型。第二部分介绍内容提供器的相关知识,首先简单地介绍了内容提供器,然后讲解如何创建内容提供器以及如何使用内容提供器访问其他程序共享的数据。第三部分讲解三种获取 Android 系统中动态权限的方法。

课后练习

【微信扫码】
第 7 章课后练习

第 8 章

Android 的数据存储

8.1 文件存储

在 Android 应用开发过程中，经常需要对文件进行的各种操作，如创建文件、读写文件等 Android 系统对文件的 I/O 操作上主要还是使用 Java 的 java.io.FileInputStream 类来完成。

Android 系统下的文件，可以分为以下两类。第一类是私有文件，即 Android 应用自己创建的文件信息，如一个 Android 游戏用来记录用户积分的文件。Android 平台中，安装好的每个应用都有属于自己的一个私有目录，所有用于该应用的数据、文件等均放在该目录中。其他应用不能随便读取写入操作（可以使用 Content Provider）访问。只有应用程序本身能对其进行操作。这个私有目录位于 Android 系统的/data/data/<应用程序包>/目录中。第二类是共享文件，如存储在 SD 卡上的文件信息，应用程序还可以对 Android 设备上的 SD 卡进行读写操作，SD 卡的路径为/sdcard/，这类文件任何 Android 应用都可以访问。

8.1.1 数据存储到文件

Android 系统提供几个方法用于文件的读写操作，以简化使用标准的 I/O 步骤。

表 8-1 Android 文件系统的读写方法

方法名	描 述
openFileInput(String name)	返回 FileInputStraeam 输入流
openFileOutput(String name, int mode)	返回 FileOutputStream 输出流
deleteFile(String name)	删除指定的文件
fileList()	返回与应用程包相关的所有私有的文件名

表 8-1 中的 name 表示要操作的文件名，不能包含路径分隔符。

使用上述方法时要注意：只支持操作当前 Android 应用程序下的文件，即应用的私有目录。传入的文件名不能带有任何的路径信息，只需要传入文件名即可。

Mode 为操作模式，默认使用 0 或 MODE_PRIVATE 。创建文件时，如果指定的文件不存在，则 Android 会创建文件，而对于存在的文件，默认使用覆盖私有模式（Context.MODE_PRIVATE ）对文件进行写操作。操作成功后返回 FileOutputStream 对象，供下一步操作使用。

如果想以增量方式写入存在的文件，需要指定输出模式（context.MODE_APPEND）。

如果打算让其他的应用程序访问输出文件，可以设置输出模式为 Content.MODE_

WORLD_READABLE 或可读写 Content.MODE_WORLD_WRITEABLE。如果参数指定的文件因权限不足或者是目录不存在等因素不能被创建，不能打开时，将抛出 FileNotFoundException。

在安卓模拟器中如果需要查找文件所在位置，可以直接打开 DDMS 视图，在/data/data/< package name >/files/文件夹中发现输出的文件。下面学习如何进行文件的存储。

【微信扫码】
案例 8－1
相关文件

案例 8－1：Android 本地文件的存储

新建一个 Android 项目，命名为 FS_IStorage_FileInOutputStream_Scanner。

步骤 1：新建一个布局文件，里面放置一个按钮和 TexTview，代码如下。

```
<Button android:id="@+id/bntShow"   android:text="显示信息" ……  />
<TextView android:id="@+id/msg"   android:textColor="#000000" ……  />
```

步骤 2：在 MainActivity.java 中修改代码，实现信息存储功能，核心代码如下。

```
public class MainActivity extends Activity {
    ………… //声明控件
    @Override
    protected void onCreate(Bundle savedInstanceState) {
        …………;
        FileOutputStream output = null ; //接收文件输出对象
        try { //采用文件流方式将信息存储到系统文件中
            output = super.openFileOutput(FILENAME,Activity.MODE_PRIVATE) ;
        } catch (FileNotFoundException e) { e.printStackTrace();  }
        PrintStream out = new PrintStream(output);
        out.println("姓名:安卓开发课程;") ;out.println("年龄:10;")
        out.println("地址:常熟理工学院") ;out.close() ; //资源一定要关闭
        bntShow = (Button) findViewById(R.id.bntShow);
        …… //跳转到信息显示界面
        bntShow.setOnClickListener(new View.OnClickListener() { …… } }
```

步骤 3：新建一个类，命名为 LoadActivity.java，用来输出文件信息。

```
public class LoadTxtActivity extends Activity {
   private static final String FILENAME = "fsfileoutputstream.txt"; //设置文件名称
   private TextView msg = null ;       //设置文本组件
   @Override
   protected void onCreate(Bundle savedInstanceState) {
      …………;
      FileInputStream input = null ;
```

```
try {…… //使用输入流获得之前存储的信息,然后输入显示
    input = super.openFileInput(FILENAME) ;
} catch (FileNotFoundException e) { e.printStackTrace();  }
Scanner scan = new Scanner(input) ;
while(scan.hasNext()){this.msg.Append(scan.next() + "\n");} scan.close() ; }}
```

运行程序,点击按钮,跳转之后可以看到效果,如图 8-1 所示。

在 IDE 右侧的 Device File Explorer 中点击查看文件目录,如图 8-2 所示。

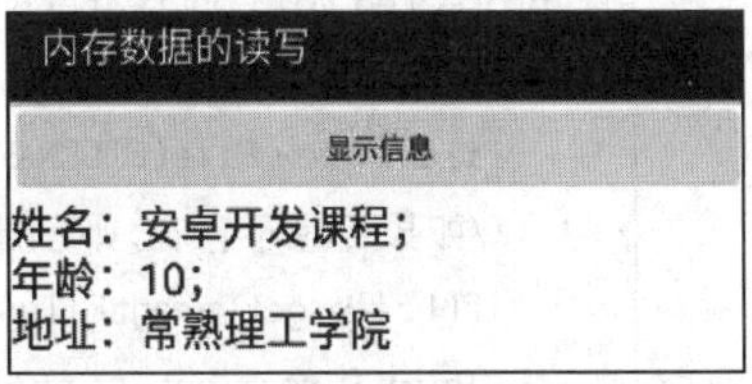

图 8-1　采用输入输出流存储信息

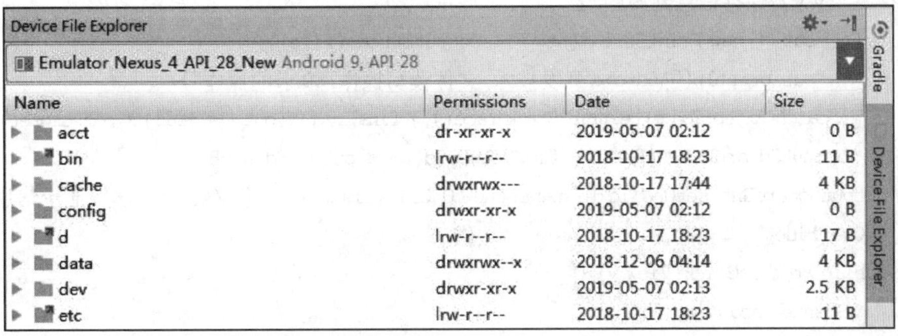

图 8-2　本地文件目录

点击 data 文件夹,查看存储的文件,如图 8-3 所示。

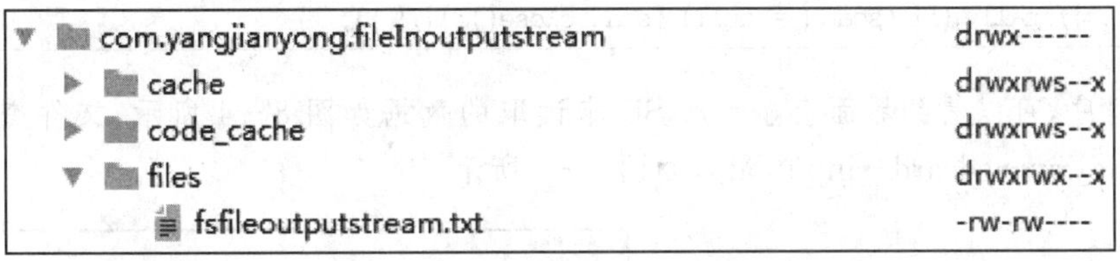

图 8-3　本地存储文件信息位置

文件所在的位置是 data→data→ApplicationId "com. yangjianyong. fileInoutputstream"→files→fsfileoutputstream.txt。

8.1.2　从 SD 卡中读取数据

如果想将文件存储到外置的 SD 卡中,操作之前先要获得操作权限:< uses-permission android:name ="android.permission.WRITE_EXTERNAL_STORAGE" / >,下面我们通过一个案例来学习。

案例 8-2:SD 卡的文件的读写

新建一个 Android 项目,命名为 FS_EStorage_IO_SdCard。

步骤 1:编写布局文件,放置一个 Button 和 TextView,代码略。

步骤 2:编写 Activity.java 文件。

【微信扫码】
案例 8-2
相关文件

```
public class InOutputSdcardActivity extends Activity {
    private Button outputSdcardBnt;private TextView msg = null ;private File file;
    //设置文件名称
    private static final String FILENAME = " /mnt /sdcard /inputdata /myyjy.txt";
    @Override
    protected void onCreate(Bundle savedInstanceState) {
      …………
      file = new File(FILENAME); //定义要操作的文件和创建父文件夹路径
      //如果目录不存在,则创建父目录
      if(!file.getParentFile().exists()) {file.getParentFile().mkdirs();}
       PrintStream out = null;
       try { //保存信息到 SD 卡上
          out = new PrintStream(new FileOutputStream(file));
          out.println("Android 开发技术,文件读写操作,SD 卡的操作");
       } catch(Exception e){e.printStackTrace();} finally {if(out ! = null) { out.close();}}
       outputSdcardBnt = (Button) findViewById(R.id.outputSdcardBnt);
    outputSdcardBnt.setOnClickListener(new OnClickListener() { //从 SD 卡读取显示信息
    @Override
    public void onClick(View v) {
        Scanner scan = null;
      try {scan = new Scanner(new FileInputStream(file));
      while(scan.hasNext()) {InOutputSdcardActivity.this.msg.Append(scan.next() + "\n");}
        } catch (FileNotFoundException e) {e.printStackTrace();
        }finally {if (scan ! = null) {scan.close();}} } }); }}
```

运行程序,可以看到界面上显示从 SD 卡读取的数据如图 8-4 所示,文件查看位置在 mnt→media_rw→sdcard→inputdata,如图 8-5 所示。

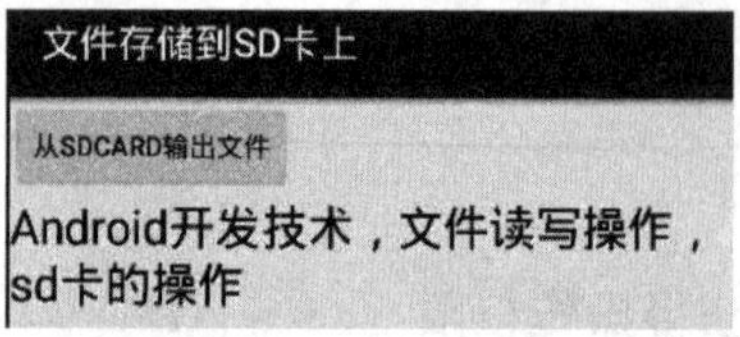

图 8-4　SD 卡文件信息的显示

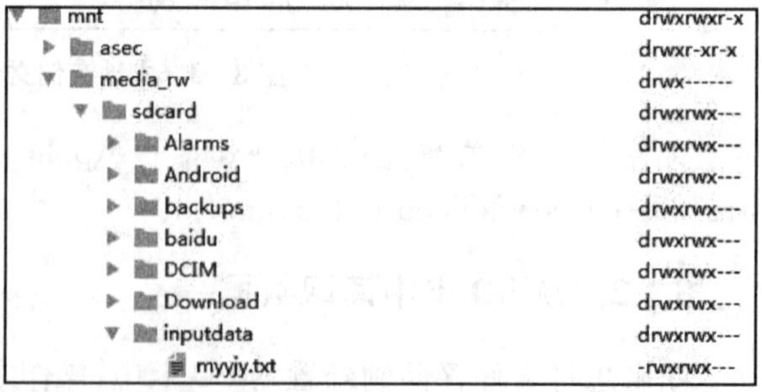

图 8-5　文件在 SD 上存储的位置

8.1.3　使用 Environment 类存储 SD 文件

上面虽然使用 I/O 流完成了文件的保存,但存在一个问题,文件的路径采用的是硬编码的方式设置,那么就有可能因为 SD 卡不存在而出现错误,最好的做法是在保存之前先判断 SD 卡是否存在,如果存在则保存,如果不存在则提示用户“SD 卡不存在”,无法保存。这个判断的功能需要使用 android. os. Environment 类取得目录的信息。调用方法 Environment.getExternalStorageState(). equals(Environment. MEDIA_MOUNTED)来测试运行环境。SD 卡上的文件或者数据是非私有数据,可以用 Java 的 I/O API 直接打开 SD 卡在 Android 系统中的映射目录,一般默认映射到/sdcard,因此可以直接使用/sdcard,当然也可以通过方法 Environment.getExternalStorageDirectory()获得目标的效果。

表 8-2 Environment 定义的常量及方法

常量及方法	描述
public static final String MEDIA_MOUNTED	扩展存储设允许进行读/写访问
public static final String MEDIA_CHECKING	扩展存储设处于检查状态
public static final String MEDIA_MOUNTED_READ_ONLY	扩展存储设处于只读状态
public static final String MEDIA_REMOVED	扩展存储设不存在
public static final String MEDIA_UNMOUNTED	没有找到扩展存储设
public static File getDataDirectory()	取得 data 目录
public static File getDownloadCacheDirectory()	取得下载的缓存目录
public static File getExternalStorageDirectory()	取得扩展的存储目录
public static String getExternalStorageState()	取得扩展存储设备的状态
public static File getRootDirectory()	取得 root 目录
public static boolean isExternalStorageRemovable()	判断扩展的存储目录是否被删除

SD 卡的容量是通过 StatFs 类来获取的,Android 系统是基于 Linux 系统,所以存储空间是以数据块的形式划分的,SD 卡容量需要通过数据块乘以每个数据块的大小而得到。下面通过一个案例来学习。

案例 8-3:使用 Environment 类读取 SD 卡上的数据

【微信扫码】
案例 8-3
相关文件

步骤 1:在项目 FS_EStorage_IO_SdCard 中新建布局文件 environment_activity_main.xml,放置一个 Button 和 TextView,用于 Environment 类读取 SD 卡,代码略。

步骤 2:新建 Activity 类,命名为 EnvironmentActivity。

```
public class EnvironmentActivity extends Activity {
  private static final String FILENAME = "envir.txt" ;   //设置文件名称
  private static final String DIR = "iosddir" ;  //设置操作文件夹的名称
  private TextView textInfo; private Button bntShow;private File file;
  @Override
  protected void onCreate(Bundle savedInstanceState) {
   ............
  if(Environment.getExternalStorageState().equals(Environment.MEDIA_MOUNTED)) {
     file = new File(Environment.getExternalStorageDirectory() + File.separator + DIR +
File.separator + FILENAME);
    if (!file.getParentFile().exists()) {file.getParentFile().mkdirs();}
```

```
    PrintStream out = null;
      try {out = new PrintStream(new FileOutputStream(file));
          out.println("Android 开发技术,文件读写操作,使用 Environment 类操作");
      }catch (Exception e) { e.printStackTrace();}finally {if (out ! = null) {out.close();}}
      }else {Toast.makeText(this, "保存失败,SD 卡不存在!",Toast.LENGTH_LONG).show();}
   bntShow.setOnClickListener(new OnClickListenerImpl()); }
   public class  OnClickListenerImpl implements View.OnClickListener{
       @Override
       public void onClick(View v) {
          try {Scanner scanner = new Scanner(new FileInputStream(file) );
             while(scanner.hasNext()) {EnvironmentActivity.this.textInfo.Append(scanner.
next() + "\n");} } catch (FileNotFoundException e) { e.printStackTrace();  } }   }}
```

运行程序,效果如图 8-6 所示。

文件查看位置在 mnt→media_rw→sdcard→iosddir,如图 8-7 所示。

图 8-6　Environment 类读取文件

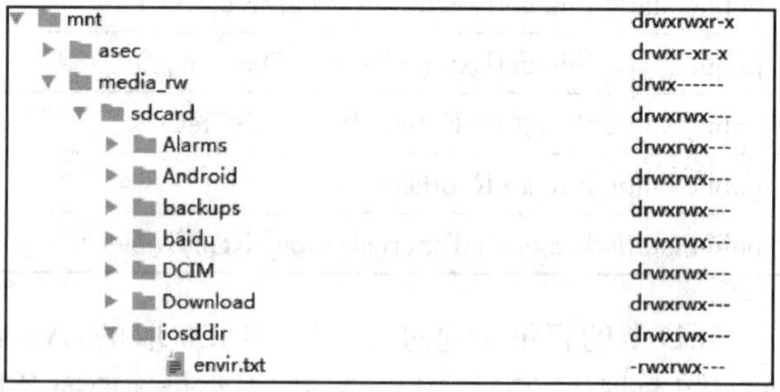

图 8-7　使用 Environment 类读取的文件位置

8.2　SharedPreferences

SharedPreferences 常用于存储应用程序中的一些配置信息,具有操作简单,读写方便的优势。但它只支持对一些简单的数据类型进行存储,支持的数据有字符串类型 string,整数类型 int,类型布尔 boolean,浮点数类型 float,长整数类型 long。

SharedPreferences 在存储机制上使用键值对的方式对数据进行存储,并提供了相应的 put/get 方法用于存储及读写操作。其存储和读写过程实际上就是一个对文件操作的过程。

在保存数据时,SharedPreferences 会在当前应用程序的私有目录(/data/data/包名/shared_Prefs)下面创建一个 XML 文件,其文件名由用户定义。如果用户没有指定文件名,系统将使用系统默认的<包名>_Preferences.xml 的文件名。

Android 中的 SharedPreferences 分为私有和公有。所谓的私有指的是针对某个 Activity 的 SharedPreferences,只有创建它的 Activity 才能访问并操作它,其他的 Activity 不能对其进行操作。公有的 SharedPreferences 则可以被应用中的多个 Activity 进行访问操作,只需要在操作时提供相应的 SharedPreferences 名字即可。

在 Android 中使用 SharedPreferences 非常简单,只需要几步便可以对其进行存储或读取。

调用 Context 对象中的 getPreferences 或 getSharedPreferences 方法获得对象，用获得的 SharedPreferences 对象操作 Preferences。

8.2.1　SharedPreferences 接口的常用方法

SharedPreferences 提供了一些基础的信息保存功能，所有的信息都是按照“key＝value”的形式进行保存的，但是 android.content.SharedPreferences 接口所保存的信息只能是一些基本的数据类型，例如：字符串、整型、布尔型等。

表 8－3　SharedPreferences 接口的常用方法

方　法	描　述
public abstract SharedPreferences.Editor edit()	使其处于可编辑状态
public abstract boolean contains(String key)	判断某一个 key 是否存在
public abstract Map < String, ? > getAll()	取出全部的数据
public abstract boolean getBoolean(String key, boolean defValue)	取出 boolean 型数据，并指定默认值
public abstract float getFloat(String key, float defValue)	取出 float 型数据，并指定默认值
public abstract int getInt(String key, int defValue)	取出 int 型数据，并指定默认值
public abstract long getLong(String key, long defValue)	取出 long 型数据，并指定默认值
public abstract String getString(String key, String defValue)	取出 string 型数据，并指定默认值

要想进行数据的写入，必须首先通过 SharedPreferences 类所提供的 edit 方法才可以让其处于可编辑的操作状态，此方法返回的对象类型是 android.content.SharedPreferences.Editor 接口实例。

表 8－4　SharedPreferences.Editor 接口的常用方法

方　法	描　述
public abstract SharedPreferences.Editor clear()	清除所有的数据
public abstract boolean commit()	提交更新的数据
public abstract SharedPreferences.Editor putBoolean(String key,boolean value)	保存 boolean 型数据
public abstract SharedPreferences.Editor putFloat(String key, float value)	保存 float 型数据
public abstract SharedPreferences.Editor putInt(String key, int value)	保存 int 型数据
public abstract SharedPreferences.Editor putLong(String key, long value)	保存 long 型数据
public abstract SharedPreferences.Editor putString(String key, String value)	保存 string 型数据
public abstract SharedPreferences.Editor remove(String key)	删除指定 key 数据

8.2.2 SharedPreferences 的数据读取方法

SharedPreferences 读取数据比较简单，直接调用相关的 get 方法即可。Android 提供了多个继承自 Preferences 的子类，例如复选框、列表、输入框 EditText 等 Preferences 实现类，使用这些实现类，可以创建各种不同类型的 Preferences，步骤如下。

(1) 调用 SharePreferences 对象的 edit 方法获得 SharePreferences.Editor 对象，此对象用于对 Preferences 进行修改。

(2) 调用 SharePreferences.Editor 中的 put 方法进行数据的修改。

(3) 调用 SharePreferences 的 commit 方法对所修改或者保存的数据进行提交操作。

下面我们将学习如何使用 SharedPreferences 来读取数据。

【微信扫码】案例 8－4 相关文件

案例 8－4：SharedPreferences 的使用

新建一个 Android 项目，命名为 DS_Teach_SharedPreferences。

步骤 1：编写布局文件。

```
<LinearLayout android:background="@color/bg" android:orientation="vertical">
    <RelativeLayout android:background="@drawable/top_title_bg" ...... >
        <ImageView android:id="@+id/login_cancel"android:onClick="onBack" ...... />
        <TextView android:text="登录" ...... />
        <Button android:id="@+id/login_register"android:onClick="onRegister"
                android:text="注册"...... />
  </RelativeLayout>
  <RelativeLayout android:background="@drawable/item_bg_1">
     <TextView android:id="@+id/email1" android:text="账号:" ...... />
     <EditText android:id="@+id/et_username" android:inputType="textEmailAddress"
               android:singleLine="true" ...... />
  </RelativeLayout>
  <RelativeLayout android:background="@drawable/item_bg_3" ...... >
     <TextView android:id="@+id/email"android:text="密码:" ...... />
     <EditText android:id="@+id/et_password" android:inputType="textPassword"
...... />
  </RelativeLayout>
  <TextView android:visibility="invisible" ...... />
  <TextView android:id="@+id/tv_findPwd"  android:text="忘记密码?" ...... />
  <TextView android:visibility="invisible" ...... />
  <LinearLayout android:id="@+id/rl_login" android:orientation="horizontal" ...... >
     <Button android:onClick="onLogin" android:text="保存用户信息" ...... />
     <Button android:onClick="onShow" android:text="显示保存信息" ...... />
  </LinearLayout>
</LinearLayout>
```

步骤 2：编写 MainActivity 代码。

```
public class MainActivity extends AppCompatActivity {
    …… //声明密码、用户名、用户名密码
    SharedPreferences mySharedPreferences; //声明信息保存类
    @Override
    protected void onCreate(Bundle savedInstanceState) {
        …… //获取布局文件、初始化和用户名密码控件 }
    …… //返回、注册、忘记密码按钮事件
    public void onLogin(View view){ //登录实现
    …… //获取用户名字和密码输入框的内容
    if (TextUtils.isEmpty(username) || TextUtils.isEmpty(password)) {
      Toast.makeText(this, "用户名或密码不可为空", Toast.LENGTH_SHORT).show();
      }else {
      mySharedPreferences = getSharedPreferences ( " myuserinfo ", MainActivity. MODE _
PRIVATE); //实例化 SharedPreferences 对象
     //实例化 SharedPreferences.Editor 对象(第二步)
    SharedPreferences.Editor editor = mySharedPreferences.edit();
    editor.putString("username", username); //用 putString 的方法保存数据
    editor.putString("pwd",password);
    editor.commit(); //提交当前数据
    Toast.makeText(this, "信息保存成功!", Toast.LENGTH_SHORT).show(); } }
    public void onShow(View v) { //显示信息
      mySharedPreferences = getSharedPreferences ( " myuserinfo ", MainActivity. MODE _
PRIVATE); //实例化 SharedPreferences 对象
     //使用 getString 方法获得 value,注意第 2 个参数是 value 的默认值
    String username = mySharedPreferences.getString("username", "没有保存的数据");
    String pwd = mySharedPreferences.getString("pwd", "没有保存的数据");
    AlertDialog.Builder builder = new AlertDialog.Builder(MainActivity.this);
        builder.setTitle("显示保存的用户信息");
        builder.setMessage("用户名:" + username + "\n" + "密码:" + pwd);
        builder.setCancelable(false);
       builder.setPositiveButton("知道了!",new DialogInterface.OnClickListener(){
        public void onClick(DialogInterface dialog, int which) {
dialog.cancel();} });
        builder.create().show(); }}
```

运行程序，在界面上输入信息，点击“保存用户信息”，如图 8－8 所示。

当点击“显示保存信息”之后，保存的信息将显示，效果如图 8－9 所示。

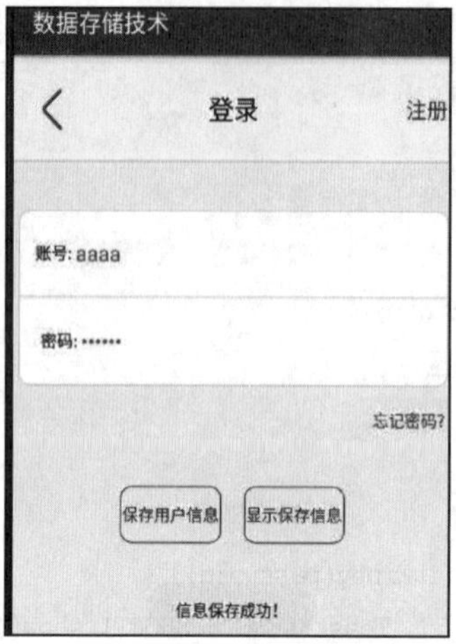

图 8-8　输入信息界面

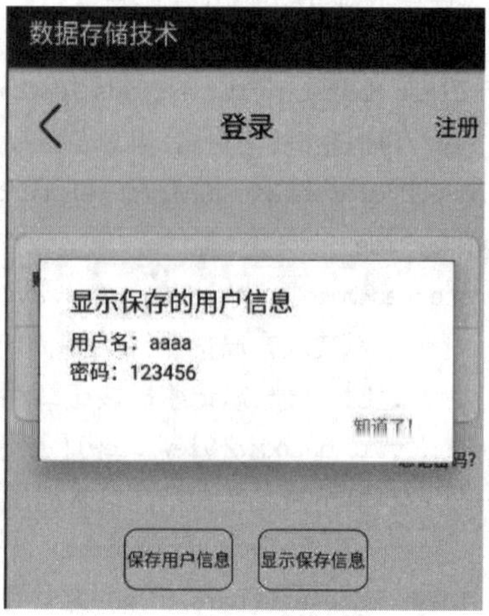

图 8-9　显示信息界面

小结

本章主要讲解了 Android 中的数据存储，首先介绍 Android 中常见的数据存储方式，接着讲解文件存储以及读取技术，最后讲解 SharedPreferences。数据存储是 Android 开发中非常重要的内容，每个应用程序基本上都会涉及数据存储，因此要求初学者必须熟练掌握本章知识。

课后练习

【微信扫码】
第 8 章课后练习

第 9 章

线程和服务

在 Android 程序中，如果需要执行那些不需要和用户交互而且长期运行的任务，就会使用服务。服务的运行不依赖于任何的用户界面，即使用户切换到后台，或者打开另外一个应用程序，服务仍能够继续保持运行。但是服务不是运行在一个独立的进程中，而是依赖于创建服务所在的应用程序。当某个应用程序被杀掉的时候，所有依赖于该进程的服务也会停止。

应用程序默认运行在主线程中。如果有一些自定义服务需要运行，那么应该创建子线程执行自定义的任务。一般不允许在主线程中执行这类任务，否则可能会出现主线程被堵塞的情况。例如，我们发出一条网络请求，由于网络运行的环境比较复杂，未必能够迅速收到响应，如果把这种操作放在主线程中，就会造成堵塞，影响正常的软件使用，就需要把这些操作放在子线程中去执行。为了学习 Android 的多线程，先来介绍下 Java 的多线程知识。

9.1 Java 的多线程

Java 中创建线程有两种方式：一种是继承 java.lang.Thread 类，另一种是实现 Runnable 接口。下面对这两种创建方式分别进行介绍。

9.1.1 Thread 类实现的多线程

Java 的线程类 Thread，可以通过继承 Thread 类，覆盖其 run()方法创建线程类。在实现上可以首先新建一个继承 Thread 的新线程类，并重写父类中的 run()方法。然后 new 出这个新线程类的实例，调用 start()方法，就会在子线程中运行了。

案例 9－1：使用 Thread 类实现的 Java 多线程

在 Eclipse 中新建 Java 项目命名为 Thread_Teach_Java，在 src 中新建一个 Java 程序命名为 ThreadDemo。

【微信扫码】
案例 9－1
相关文件

步骤 1：新建一个线程类 PrintAThread 继承于 Thread，用于打印字母 A，代码如下。

```
class PrintAThread extends Thread {
    public void run() {
        for (int i = 0; i <= 5; i++) {System.out.print("-A-");
            try { sleep(500); //线程休眠 500 毫秒
                } catch (InterruptedException e) { }} } }
```

程序的功能是打印字母 A。

步骤 2：新建线程类 PrintBThread，用于打印字母 B，代码参照 PrintBThread。

步骤 3：编写 main 函数，用于启动两个线程。

```
public class ThreadDemo {
    public static void main(String[ ] args) {
        PrintAThread printa = new PrintAThread();
        PrintBThread printb = new PrintBThread();
        printa. start(); printb. start();  } }
```

-A--B--A--B--A--B--B--A--A--B--A--B-

图 9－1　继承 Thread 类的线程运行结果

在 main 函数中运行结果如图 9－1 所示。从程序的运行结果中可以发现，现在的两个线程对象是交错运行的，哪个线程对象抢到了 CPU 资源，哪个线程就可以运行，所以程序每次的运行结果肯定是不一样的，在线程启动时虽然调用的是 start()方法，但实际上调用的却是 run()方法定义的主体。

9.1.2　Runnable 接口实现的多线程

使用继承 Thread 的方法，在某些需要继承其他类的情况下就不适用了。在这种情况会选择使用 Runnable 接口方式来实现。下面介绍它的使用。

案例 9－2：使用 Runnable 接口实现的多线程

在项目 Thread_Teach_Java 的源代码 src 中，新建一个 Java 程序，命名为 ThreadRunnable，编写代码如下。

【微信扫码】
案例 9－2
相关文件

步骤 1：编写打印字母 A 的线程，继承于接口 Runnable。

```
class PrintARunnable implements Runnable  {
    public void run() {
        for (int i = 0; i <= 5; i++) {System. out. print(" - A - ");
          try {Thread. sleep(500);}catch (InterruptedException e){e. printStackTrace();}}}}
```

步骤 2：编写打印字母 B 的线程，继承于接口 Runnable，代码参照 PrintARunnable。

步骤 3：编写 main 函数，用于启动两个线程。

```
public class ThreadRunnable {
    public static void main(String[ ] args) {
        PrintARunnable pa = new PrintARunnable(); //创建线程 left
        PrintBRunnable pb = new PrintBRunnable(); //创建线程 right
        Thread threadpa = new Thread(pa);Thread  threadpb = new Thread(pb);
        threadpa. start(); threadpb. start(); }}
```

要想启动一个多线程必须要使用 start()方法完成，如果继承了 Thread 类，则可以直接从 Thread 类中使用 startQ 方法，但是现在实现的是 Runnable 接口并没有 start()方法的定义，那么该如何启动多线程呢？需要依靠 Thread 类完成启动，在 Thread 类中提供了 public Thread(Runnable target)和 public Thread(Runnabletarget，String name)两个构造方法。

这两个构造方法都可以接收 Runnable 的子类实例对象，可以依靠此点启动多线程。在 main 函数中启动两个线程，运行结果如图 9－2 所示。分析运行结果，我们发现实现效果和继承 Thread 类的 ThreadDemo 效果相同。

```
-A--B--B--A--B--A--B--A--A--B--B--A-
```

图 9－2　继承 Runnable 接口的线程运行结果

通过 Thread 类和 Runnable 接口都可以实现多线程。但其实在 Thread 类中的 run() 方法调用的是 Runnable 接口中的 run()方法。线程使用上使用 Runnable 接口相对于继承 Thread 类来说，有如下显著的优势。

(1) 如果一个类继承 Thread 类，不适合多个线程共享资源。而采用 Runnable 接口，可以方便地实现资源共享，适合多个相同程序代码的线程去处理同一资源的情况。

(2) 可以避免由于 Java 的单继承特性带来的局限。

(3) 增强了程序的健壮性，代码能够被多个线程共享，代码与数据是独立的。

所以，在开发中建议读者使用 Runnable 接口实现多线程。有关 Java 多线程的更多知识，请参考其他 Java 的技术书籍，这里不再展开论述。

9.2　Android 的多线程

Android 中的 UI 线程也是不安全的，如果需要更新应用程序中的 UI 组件，则必须在主线程中进行，否则会出现问题。如果必须在子线程中更新 UI 组件，可以使用 Android 提供的一套异步消息处理机制。

9.2.1　子线程中更新 UI 问题

案例 9－3：Android 程序中使用子线程更新 UI

新建一个 Android 项目，命名为 Thead_Teach_UpdateUI。在本案例中，使用三种方法实现子线程中更新 UI 的方法。

【微信扫码】
案例 9－3
相关文件

步骤 1：编写布局文件，放置几个按钮，用来显示不同方式的更新控件内容的方法。

```
< TextView android:id = "@ + id /mytext"……  />
< Button android:id = "@ + id /bnt_change"    android:text = "直接改变文本内容"…… />
< Button android:text = "在线程中改变文本内容"…… />
< Button android:text = "使用 handler 中改变文本内容"…… />
< Button android:text = "使用 message 改变文本内容"……  />
```

步骤 2：实现"直接改变文本内容"按钮功能，效果如图 9－3 所示。

```
public class MainActivity extends Activity {
    …… //按钮和文本框声明
  @Override
  protected void onCreate(Bundle savedInstanceState) {
```

```
        ............
        bnt_change.setOnClickListener(new OnClickListenerImpl());
        bnt_thead_change.setOnClickListener(new OnClickListenerImpl()); }
private class  OnClickListenerImpl implements View.OnClickListener{
    @Override
      public void onClick(View v) {
        switch (v.getId()){ //通过 ID 值来区分不同的按钮
         case R.id.bnt_change:mytext.setText("直接改变的内容,没有使用线程");break;}}
```

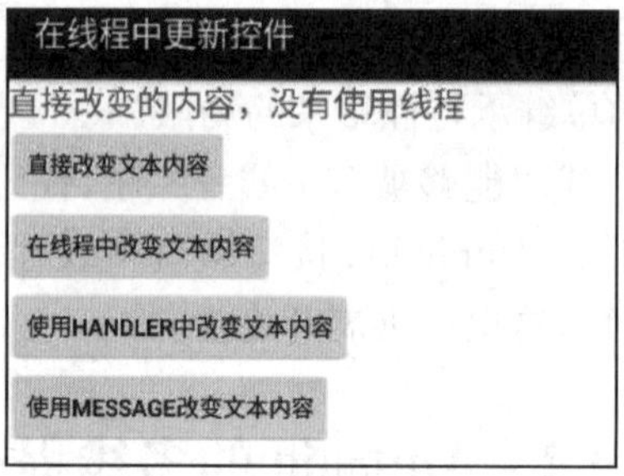

图 9-3　实现直接改变文本内容

可以看到这里没有使用线程，动态更新 TextView 的内容也成功了，但是在原则上是不可以这样做的。Android 不允许在主线程中直接修改控件内容，这样会带来隐患。

9.2.2　异步消息处理机制

Android 的异步消息处理机制包含了四个部分：Message、Handler、MessageQueue 和 Looper。每个部分用途如下。

(1) Message：在线程之间传递的消息，可以在内部携带少量的信息。

(2) Handler：处理者，用于发送和处理消息。Handler 的 sendMessage()方法用来发送消息，Handler 的 handleMessage()方法用来处理消息。

Handler 和 Message 是 Android 系统中各线程之间的一种信息通信机制。

Android 系统本身是遵循单线程模型的，当程序启动时，系统会同时启动一个相应的主线程负责处理这个程序中与 UI 相关的事件。当遇到一些比较耗时的操作时，应该建立新的子线程去执行，但是如果用子线程来更新 UI 对象，则会遇到异常 CalledFromWrongThreadException。

正确的做法是在子线程中发出 Message，然后由 Handler 获取到之后，进行 UI 操作。这也就表示可以将 Handler 理解为是 Message 的主要处理者，负责其发送与执行处理。

(3) MessageQueue：消息队列，主要用于存放所有通过 Handler 发送的消息。这些消息一直在队列中等待被处理。每个线程只会有一个 MessageQueue 对象。

(4) Looper：是每个线程中的 MessageQueue 的管家。Looper 的 loop()方法被调用之后，就会进入到一个无限的循环中，当发现 MessageQueue 中存在一条消息就会将其取出来，并传递到 Handler 的 handleMessage()方法中。每个线程也只会有一个 Looper 对象。

异步消息的处理机制流程如下。

(1) 在主线程中创建一个 Handler 对象，重写它的 handleMessage()方法。

(2) 当子线程需要进行 UI 更新操作时，创建一个 Message 对象，并通过 Handler 将这条消息传递出去。然后这条消息会被添加到 MessageQueue 的队列中等待处理。

(3) Looper 会一直尝试从 MessageQueue 中取出待处理的消息，然后转发到 Handler 的 handleMessage()方法中。

(4) Handler 在主线程中运行，它的方法也会在主线程中运行。这样 Message 就从子线程辗转到主线程中，可以对 UI 进行更新。

步骤 3：使用 runOnUiThread 在子线程中更新 UI，效果如图 9－4 所示，在 switch 中添加。

```
case R.id.bnt_thead_change:new Thread(new Runnable() {
  @Override
  public void run() { runOnUiThread(new Runnable() {
    @Override
    public void run(){mytext.setText("在线程中使用 runOnUiThread 改变的内容");
}});}}).start();break;
```

步骤 4：使用 Handler 在子线程中更新 UI，如图 9－5 所示，在 switch 中继续添加。

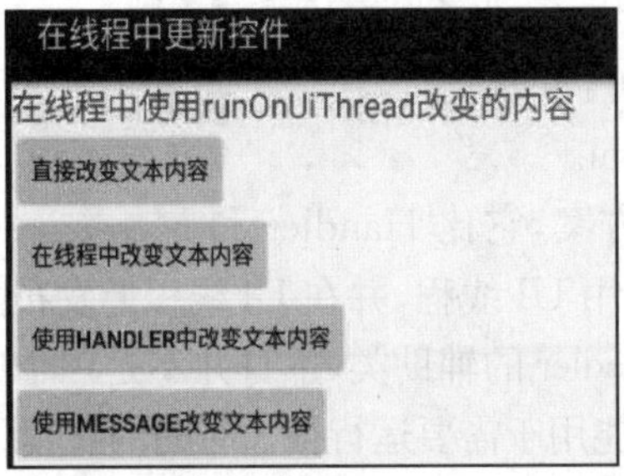

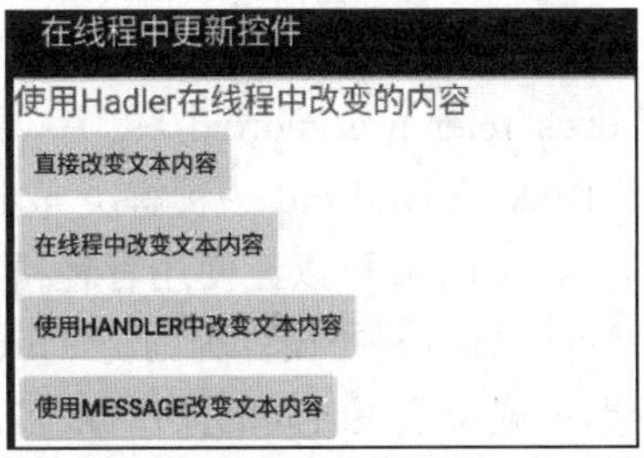

图 9－4　使用 runOnUiThread 在子线程中更新 UI　图 9－5　使用 Handler 在子线程中更新 UI

```
case R.id.bnt_handler_change:new Thread(new Runnable() {
      @Override
      public void run() {handler.post(new Runnable() {
       @Override
       public void run(){mytext.setText("使用 Hadler 在线程中改变的内容");}});
          }}).start();break;
```

步骤 5：使用 Message 在子线程中更新 UI，在 switch 中继续添加。

```
case R.id.bnt_message_change:new Thread(new Runnable(){
      @Override
      public void run() {
        String result = "使用 message 在线程中进行修改";
        Message message = new Message();
        message.obj = result; //设置数据
        handlerMessage.sendMessage(message); //发送消息
        } }).start();break;
```

按钮 bnt_message_change 使用典型的异步更新模式。在线程的 run 方法中，更新之前首先创建一个 Message 对象，然后通过 sendMessage 将这条消息发布出去。消息被保存在 MessageQueue 中，通过 Looper 取出，交给 Handler 进行处理。

```
private Handler handlerMessage = new Handler(){
   @Override
   public void handleMessage(Message msg) {
       super.handleMessage(msg); //接收消息
       String result = (String) msg.obj; //处理消息
     mytext.setText(result);}};
```

在主线程中创建的 Handler，就会在主线程进行后续处理。通过 handleMessage 接收 Message 发布的消息，然后进行更新。

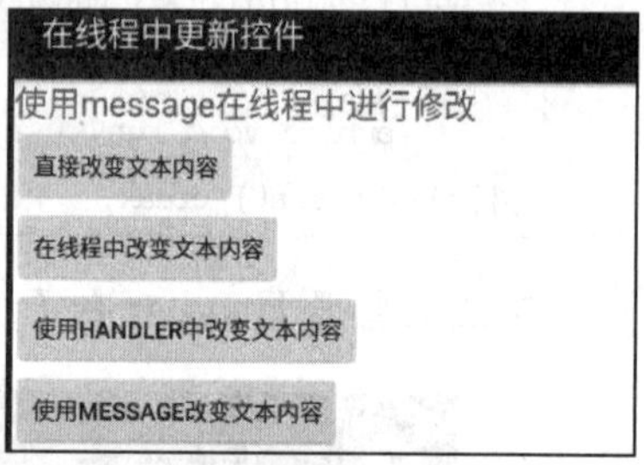

图 9－6　使用 Message 在子线程中更新 UI

9.2.3　AsyncTask 的使用

针对子线程中对 UI 更新的操作，Android 还提供了一些更好的解决方法，比如 AsyncTask。API 所在位置：SDK 安装目录/docs/reference/android/os/ AsyncTask.html。

AsyncTask 是 Android 一种解决异步加载的方案，它比 Handler 和 Message 的方式更加轻量级，AsyncTask 可以让使用者轻松正确地使用 UI 线程，并在 UI 线程上发布操作的结果。但是 AsyncTask 是被设计为围绕 Thread 和 Handler 的辅助类，本身并不是一个通用的线程处理框架，一般用于短时间的操作(最多几秒)，不能用于需要运行很长时间的线程。

AsyncTask 是一个抽象类，需要使用一个子类去继承并实现类中的方法。在继承的时候，为 AsyncTask 类指定三个泛型参数。

(1) Params(传入参数)：在执行 AsyncTask 时需要传入参数，用于在后台的任务中使用。

(2) Progress(进度数据值)：是进程执行的百分比。后台任务执行时如果需要在界面上显示当前的进度，可以指定这个泛型作为进度的单位。

(3) Result(返回结果)：当任务执行完毕，需要对结果进行返回，则在这里可以指定这个泛型作为返回值的类型。

```
Class DownloadTask extends AsyncTask < void, Integer, Boolean >{}
```

void 表示在执行 AsyncTask 中不需要传入参数给后台任务；Integer 表示使用整型数据来显示进度单位；Boolean 表示使用布尔值来返回执行的结果。

在 AsyncTask 中需要重写的四个方法如下。

(1) onPreExecute()：这个方法在后台任务开始执行之前调用，执行一些初始化的工作。

(2) doInBackground(Params…)：紧跟 onPreExecute()之后执行实际的后台操作，一般执行比较耗时的操作，如需要更新实时任务进度，可以调用 publishProgress()。这个方法的所有代码都在子线程中执行，如果在 AsyncTask 的第三个参数没有设置为 void，那任务一结束，就通过 return 语句将执行的结果返回。这个方法是不能进行 UI 操作的，如果需要对

UI 进行操作，可以调用 publishProgress(Progress…)方法来完成。

(3) onProgressUpdate(Progress…)：在后台调用了 publishProgress(Progress…)后，这个方法就会被调用。它的参数是后台传递过来的，这个方法中可以对 UI 进行更新操作，利用参数的数值就可以对界面元素进行相应的更新操作。运行于 UI 线程，可更新实时进度。

(4) onPostExecute(Result)：当后台任务执行完毕，通过 return 返回时，这个方法就会被调用。返回的数据作为参数传递到此方法中，可以根据返回的数据进行 UI 的更新操作。运行于 UI 线程，其参数就是 doInBackground(Params…)的返回值。

案例 9 - 4：使用 AsycTask 执行异步任务

新建一个 Android 项目名为 Thread_Teach_AsycTask，使用 AsycTask 来执行异步任务。

【微信扫码】
案例 9 - 4
相关文件

步骤 1：在 res/layout/activity_main.xml 中编写布局文件，放置一个 Button。

```
<Button android:text = "执行异步任务" android:id = "@ + id /button" ……  />
```

步骤 2：在 src 中新建一个继承 AsyncTask 类 DownloadAsycTask，类中实现从 AsyncTask 继承而来的抽象方法。

```
class DownloadAsycTask extends AsyncTask < Void, Integer, Boolean >{
    @Override.
    protected void onPreExecute() { //后台任务执行前执行，执行初始化工作
        super.onPreExecute();
        progressDialog = new ProgressDialog(MainActivity.this); //创建进度条对话框，参数为上下文，上下文必须是 MainActivity
       rogressDialog.setProgressStyle(ProgressDialog.STYLE_HORIZONTAL); //进度条样式
       progressDialog.show(); //显示进度条对话框}
   @Override
   protected Boolean doInBackground(Void... params) { //执行实际的后台操作
       for(int i = 0;i < = 100;i ++ ){ //模拟后台耗时的操作，每次休息 50 毫秒
         try {Thread.sleep(50);}catch (InterruptedException e) { //发生异常执行失败
            e.printStackTrace();return false;} publishProgress(i); //调用更新方法}
            return true;}
    @Override
    protected void onProgressUpdate(Integer... values){ //对 UI 进行更新操作
        super.onProgressUpdate(values);progressDialog.setProgress(values[0]);}
   @Override
   protected void onPostExecute(Boolean aBoolean) { //后台任务执行后执行
        super.onPostExecute(aBoolean);
        progressDialog.dismiss(); //隐藏进度条
        if (aBoolean) {Toast.makeText (getApplicationContext ( ), "执行成功", Toast.LENGTH_SHORT).show(); //显示对话框}}
```

```
  @Override
  protected void onDestroy() { //销毁
    super.onDestroy();
   if(progressDialog! = null){progressDialog.dismiss();progressDialog = null;}}}
```

步骤 3:编写程序代码,调用 AsyncTask 类实现功能,运行效果如图 9-7 所示,代码如下。

```
public class MainActivity extends Activity {
  Button button;ProgressDialog progressDialog;
  @Override
  protected void onCreate(Bundle savedInstanceState) {
   …………
   button.setOnClickListener(new View.OnClickListener() {
     @Override
     public void onClick(View v) {
      new DownloadAsycTask().execute(); //创建 AsyncTask 调用 execute 执行异步操作
      } });}
```

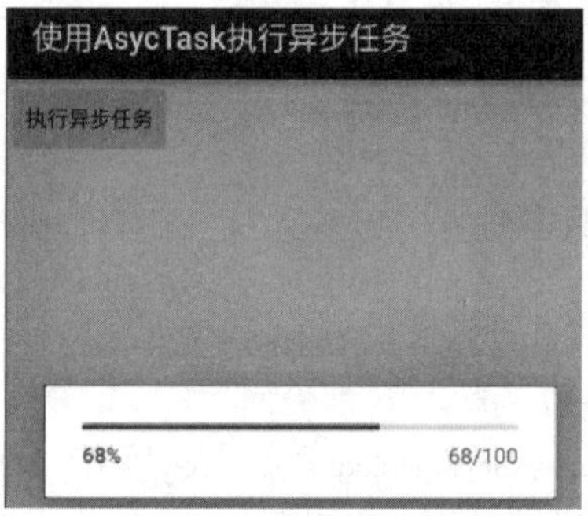

图 9-7　使用 AsyncTask 类的多线程

9.3　Service

在 Android 系统中,Service(服务)是一个重要的组成部分,是类似于 Activity 的一种组件,它没有 UI 界面,不能自己启动,也不能与用户交互,只能运行于后台。常用于那些长时间运行,且在运行的过程中又不需要与用户进行交互的任务,例如,退出音乐播放器时继续在后台播放音乐的功能。

如果某些程序中的操作很消耗时间,可以将这些操作定义在 Service 之中,这样就可以让程序在后台运行(也可以在不显示界面的形式下运行),相当于是一个没有图形界面的 Activity 程序。当用户要执行某些操作需要进行跨进程访问操作的时候,也可以使用 Service 来完成,在开发时用户只需要继承 android.App.Service 类就可以完成 Service 程序的开发。

服务在运行过程中,不依赖外部的界面,即使当某些程序被切换到后台,或者其他的应用程序被打开，服务依然能够正常运行。服务的运行，依赖于服务被创建时所依赖的应用程序进程，当这个进程被 kill 的时候，所有依赖这个进程的服务也将被终止运行。服务默认运行在主线程中，不会自行开启线程运行，需要使用时在服务内部手动创建子线程，在其内执行具体的任务。

9.3.1 Service 的生命周期

Service 的生命周期比 Activity 简单很多，但是却需要更加关注服务如何创建和销毁，因为服务在用户不知情时就可以在后台运行。Service 的生命周期分成以下两个类。

(1) Started Service：在项目的任何位置调用 Context.startService()方法，相应的服务就会启动，接着服务无限期运行。如果要停止服务，自身必须调用 stopSelf()方法或者其他组件调用 stopService()方法。当服务停止时，系统将其销毁。

(2) Bound Service：当其他组件调用 Context.bindService()，也会启动一个服务，这个服务的连接属于持久连接。客户端通过 IBinder 接口与服务通信。如果客户端需要关闭连接，则通过 unbindService()实现。多个客户端能绑定到同一个服务并且当它们都解绑定时，系统销毁服务(服务不需要被停止)。

注意：如果对一个服务既调用了 startService()方法，又调用了 bindService()方法，则需要同时调用 stopService()和 unbindService()方法，onDestroy()方法才能被执行。

Service 生命周期中会用到回调的方法，用于根据需要提供组件绑定到服务，方法如下。

(1) onCreate：当服务第一次创建时，系统调用该方法执行一次性建立，系统调用 onStartCommand()或 onBind()方法。如果服务已经运行，该方法不被调用。

(2) onStartCommand：当其他组件，例如 Activity 调用 startService()方法请求服务启动时，系统调用该方法。一旦该方法执行，服务就启动(处于 Started 状态)并在后台无限期运行。如果开发人员实现该方法，则需要在任务完成时调用 stopSelf()或 stopService()方法停止服务(如果仅想提供绑定，则不必实现该方法)。

(3) onBind：当其他组件调用 bindService()方法想与服务绑定时(例如执行 RPC)，系统调用该方法。在该方法的实现中，开发人员必须通过返回 IBinder 提供客户端用来与服务通信的接口。该方法必须实现，但是如果不想允许绑定，则应该返回 null。

(4) onUnBind：当调用者通过 unbindService()函数取消绑定 Service 时 onUnBind()方法将被调用。如果 onUnbind 函数返回 true，表示重新绑定服务时 onRebind()函数将被调用。

(5) onDestroy：当服务不再使用并即将销毁时，系统调用该方法。服务应该实现该方法来清理诸如线程、注册监听器、接收器等资源。

当项目本身调用服务，如果是第一次启动，服务还没有被创建时，onCreate()就会首先执行，onStartCommand()也会被调用。以后每次调用服务只会调用 onStartCommand()。服务启动之后会一直保持运行状态。如果需要结束或者销毁服务，调用 stopService 或者 stopSelf()方法。每个服务只会存在一个实例，每次调用 stopService()或者 stopSelf()服务就会停止下来。

当其他组件调用服务，类似前面的描述，如果服务还没有被创建，第一次启动时，onCreate()就会先于 onBind()方法执行。调用方会获取 onBind()方法中返回的 IBinder 对象实例。这样调用方就可以一直与服务进行通信。调用 unbindService()方法，可以将服务销毁。

9.3.2 Service 的分类

Local Service(本地服务):用于应用程序内部。用于实现应用程序自己的一些耗时任务,如查询升级信息,并不占用应用程序如 Activity 所属线程,而是另开线程后台执行。

Remote Sevice(远程服务):用于 Android 系统内部的各应用程序之间。可被其他应用程序复用,如天气预报,其他应用程序不需要再写这样的服务,调用已有的即可。

实现 Service 需要继承 Service 类,其启动的方式有两种:startService()和 bindService(),相对应的结束方式分别为 stopService()和 unbindService()。

startService()方法启用服务:调用者与服务之间没有关联,即使调用者退出,服务仍然运行。bindService()方法启用服务:调用者与服务绑定一起,调用者一旦退出,服务也就终止了。

9.3.3 定义和创建 Service

Service 的定义中有两个属性:Exported 属性表示是否允许除了当前程序之外的其他程序访问这个服务;Enable 属性表示是否启用这个服务。

Service 类中唯一的抽象方法 onBind(),必须在继承的子类中实现。

服务可以从 Context.startService()和 Context.bindService()开始。当服务第一次启动的时候,onCreate ()和 onStartCommand ()都会被调用。但随后的每次调用,只会调用 onStartCommand()方法。

如果在 Activity 中对 Service 进行管理,使用 IBinder onBind (Intent intent) 方法。如果客户端没有绑定服务返回 null,通常返回 IBinder 作为 Android Interface Definition Language (Android 界面定义语言)的复杂接口。与其他应用不同,调用返回到这里的 IBinder 接口可能不会在进程的主线程上。使用过程中一般创建一个专门的 Binder 对象进行管理。

接口 IBinder 的 API 地址为:SDK 安装目录/docs/reference/android/os/IBinder.html。

类 Binder 的 API 地址为:SDK 安装目录/docs/reference/android/os/Binder.html。

下面通过一个案例来演示 Service 的使用。

案例 9-5:Service 的使用

新建一个 Android 项目,命名为 Service_Teach_ServiceDemo。我们将实现服务的创建、启动和停止,并在服务中绑定和解绑其他功能。

【微信扫码】案例 9-5 相关文件

步骤 1:在 Android Studio 中点击 File/New/Service/Service,创建一个继承 Service 的类 MyService,用于创建服务,onBind()方法是抽象方法,必须要子类实现,进行重写代码如下。

```
public class MyService extends Service {
    @Override
    public void onCreate() { //服务创建的时候调用
        super.onCreate();Log.i("MyService","-- onCreate"); }
    @Override
    //服务每次启动的时候调用
```

```
    public int onStartCommand(Intent intent, int flags, int startId) {
        Log.i("MyService","-- onStartCommand");
        return super.onStartCommand(intent, flags, startId); }
    @Override
    //服务销毁的时候调用
    public void onDestroy() {super.onDestroy();Log.i("MyService","-- onDestroy");}
  @Nullable
  @Override
  public IBinder onBind(Intent intent) { return null; }}
```

步骤 2：Service 属于应用程序的组件，每个服务都必须在 Androidmanifest.xml 中的 service 标签中进行注册才能生效。点击打开 Androidmanifest.xml，新增代码如下。

```
<service android:name = ".MyService" android:enabled = "true" android:exported = "true"  />
```

android：name=".MyService"是实现服务的 Service 子类名称。这是一个完整的类名，如果名称的第一个符号是点号(例如.RoomService)，它会增加在 manifest 标签中定义的包名。它没有默认值并且必须指定。

android：enabled="true"：服务能否被系统实例化，true 为可以，false 为不可以，默认值是 true。

android：exported="true"：其他组件能否调用服务或与其交互，true 为可以，false 为不可以。

步骤 3：在 app/res/layout/activity_main.xml 编写布局文件，设计两个按钮，用来启动和停止服务，代码如下。

```
<Button android:text = "开启服务" android:onClick = "start" android:id = "@+id/button"……  />
<Button android:text = "停止服务" android:onClick = "stop" android:id = "@+id/button2"…… />
```

步骤 4：修改 MainActivity，实现启动和停止服务功能，代码如下。

```
public class MainActivity extends AppCompatActivity {
    @Override
    protected void onCreate(Bundle savedInstanceState) { ……}
    public void start(View view) {
        Log.i("MainActivity","-----启动服务----");
        Intent intent = new Intent(this,MyService.class);startService(intent);}
    public void stop(View view) {
        Log.i("MainActivity","----停止服务----");
        Intent intent = new Intent(this,MyService.class);stopService(intent); }
```

运行程序，结果如图 9-8 所示，可以看出当启动的时候先后调用 onCreate()和 onStartCommand()，当停止的时候调用 onDestroy()。如果连续点击启动服务，运行结果如图 9-9 所示。可以发现，只有服务第一次创建的时候调用 onCreate()方法，后面就不会再调用了。

```
I/MainActivity: ----启动服务----
I/MyService: --onCreate
I/MyService: --onStartCommand
I/MainActivity: ----停止服务----
I/MyService: --onDestroy
```

图 9-8 启动和停止服务的结果

```
I/MainActivity: ----启动服务----
I/MyService: --onCreate
I/MyService: --onStartCommand
I/MainActivity: ----启动服务----
I/MyService: --onStartCommand
I/MainActivity: ----启动服务----
I/MyService: --onStartCommand
I/MainActivity: ----启动服务----
I/MyService: --onStartCommand
I/MainActivity: ----停止服务----
I/MyService: --onDestroy
```

图 9-9 连续点击启动和停止服务

步骤 5：修改 MyServices.java 代码实现绑定服务，提供一个下载功能，然后在 Activity 中可以决定何时开始下载以及随时查看下载进度。因为需要服务告诉 Activity 数据的变化，使用 onBind()方法实现，添加代码如下。

```
public class MyService extends Service {
    …… //省略步骤 1 的方法
    //新建了一个继承自 BinderDownload 的 Binder 类
    private DownloadBinder downloadBinder = new DownloadBinder();
    //在 DownloadBinder 类内部模拟了开始下载以及查看下载进度的方法
    class DownloadBinder extends Binder{
        public void startDownload(){Log.d("MyService", "开始下载");}
        public int getProgress(){Log.d("MyService", "进度变化");return 0;}}
    @Nullable
    @Override
    public IBinder onBind(Intent intent) {return downloadBinder;}}
```

步骤 6：打开 activity_main.xml 修改布局文件，添加服务绑定和解除按钮，代码如下。

```
<Button android:text="绑定服务" android:onClick="bind"    ……  />
<Button android:text="解绑服务" android:onClick="unBind" ……  />
```

步骤 7：在 MainActivity.java 中实现绑定和解除服务，添加代码如下。

```
public class MainActivity extends AppCompatActivity {
    …… //省略步骤 4 的方法
    private MyService.DownloadBinder downloadBinder;
    //创建了一个 ServiceConnection 的匿名类
    private ServiceConnection connection = new ServiceConnection(){
```

```
@Override
//重写 onServiceDisconnected()方法,Activity 与 Service 解除绑定时调用
public void onServiceDisconnected(ComponentName name){ }
@Override
//重写 onServiceConnected()方法,Activity 与 Service 成功绑定时调用
public void onServiceConnected(ComponentName name,IBinder service){
    downloadBinder = (MyService.DownloadBinder)service; //获得 DownloadBinder 实例
    downloadBinder.startDownload();downloadBinder.getProgress();} };
public void bind(View view) { //绑定服务
    Log.i("MainActivity","--绑定服务--");
    Intent bindIntent = new Intent(this, MyService.class); 构建一个 Intent 对象
    //传入 BIND AUTO CREATE 表示在活动和服务进行绑定后自动创建服务
    bindService(bindIntent, connection, BIND_AUTO_CREATE); }
public void unBind(View view) { //解绑服务,直接调用 unBindService。
    Log.i("MainActivity","--解绑服务--");
    unbindService(connection);}}
```

运行程序,结果如图 9-10 所示。

```
I/MainActivity: ----启动服务----
I/MyService: --onCreate
I/MyService: --onStartCommand
I/MainActivity: ----停止服务----
I/MyService: --onDestroy
I/MainActivity: --绑定服务--
I/MyService: --onCreate
D/MyService: 开始下载
D/MyService: 进度变化
I/MainActivity: --解绑服务--
I/MyService: --onDestroy
```

图 9-10 绑定和解绑服务的效果

9.4 IntentService

服务中的代码都是默认运行在主线程当中的,如果直接在服务里去处理一些耗时的逻辑,就很容易出现 ANR(Application Not Responding)的情况。这个时候就需要用到 Android 多线程编程的技术了,直接的做法就是每个实现的方法里开启一个子线程,在子线程中去处理那些耗时的逻辑。当线程停止的时候,调用 stopSelf()停止当前服务。

虽说这种写法并不复杂,但是当子线程比较多的时候,总会有一些程序员忘记开启线程和停止服务。为了解决这个问题,Android 专门提供了一个异步的、会自动停止的服务

IntentService 类，它是 Service 的子类，这个类很好地解决了异步的问题。

IntentService 是继承自 Service 并处理异步请求的一个类，在 IntentService 内有一个工作线程来处理耗时操作。当任务执行完，IntentService 会自动停止。

如果启动 IntentService 多次，那么每一个耗时操作会以工作队列的方式在 IntentService 的 onHandleIntent 回调方法中执行，使用串行方式依次执行，执行完成后自动结束。

IntentService 中使用的 Handler、Looper、MessageQueue 机制把消息发送到线程中去执行，所以多次启动 IntentService 不会重新创建新的服务和新的线程，只是把消息加入消息队列中等待执行，如果服务停止，会清除消息队列中的消息，后续的事件得不到执行。

案例 9－6：使用 IntentService 进行多线程处理

新建一个 Android 项目，命名为 Service_Teach_IntentService_ToSum。

在本案例中，我们实现一个连续进行计算的应用。在主程序中启动服务，并传递两个参数 a 和 b，在 IntentService 中获取参数，进行求和，并通过发送广播的形式向 UI 返回结果，然后 MainActivity 在接收到广播之后更新 UI。这里需要使用广播的相关知识。

【微信扫码】
案例 9－6
相关文件

步骤 1：在 app/res/layout/activity_main.xml 编写布局文件，放置一个按钮用于启动服务。放置一个线性布局用于显示计算的结果，代码如下。

```
< Button android:text = "发起计算请求" android:onClick = "addTask" ……  />
< LinearLayout android:id = "@ + id /ll_container" android:orientation = "vertical" ……  />
```

步骤 2：编写常量类，在 MainActivity 和服务之间需要传递数据，使用常量类，代码如下。

```
public class Constants {
    public static final String ACTION_FOR_SUM  =  "com.service.intentservicesum.action.ACTION
_FOR_SUM";
    public static final String ACTION_RESULT = "com.service.intentservicesum.action.ACTION_
RESULT";
   public static final String   EXTRA_A = "com.service.intentservicesum.extra.A";
   public static final String EXTRA_B = "com.service.intentservicesum.extra.B";
   public static final String EXTRA_RESULT = "com.service.intentservicesum.extra.RESULT;}
```

步骤 3：新建一个继承 IntentService 的类 MyIntentService 实现 IntentService 服务类的生命周期，代码如下。

```
public class MyIntentService extends IntentService {
    public static final String TAG = " ===> MyIntentService";
    public MyIntentService() {super("MyIntentService");}
    @Override
    public void onCreate() {Log.i(TAG,"onCreate()");super.onCreate();}
    …… //完善生命周期的其他方法:onStart、onStartCommand、onBind、onDestroy}
```

步骤 4:在 MyIntentService 类中重写服务类 onHandleIntent()方法。

```
@Override
protected void onHandleIntent(Intent intent) { //Intent 是 StartService 方法传递过来的
  if (intent! = null){ //当传递的 Intent 不为空时,提取 Intent 中的 action 值
    String action = intent.getAction();
    if(Constants.ACTION_FOR_SUM.equals(action)){ //传递 action 是 ACTION_FOR_SUM
    int a = intent.getIntExtra(Constants.EXTRA_A,0); //提取 a 和 b 的值,进行求和运算
    int b = intent.getIntExtra(Constants.EXTRA_B,0);
    int result = a + b;
    Log.i(TAG,"onHandleIntent():" + android.os.Process.myTid() + ":result: " + result);
    handleResult(a,b,result); //传递给 handleResult 方法}}}
```

步骤 5:在 MyIntentService 类中,新增一个方法 handleResult,功能是将运算结果封装后返回给消息接受者,代码如下。

```
private void handleResult(int a, int b, int result){
   try{Log.i(TAG,"handleResult: " + a + " + " + b + " = " + result);
     Thread.sleep(2000); //模拟计算耗时,让线程休眠 2 秒
      //action 为 ACTION_RESULT 为显示运算结果
     Intent intent = new Intent(Constants.ACTION_RESULT);
    intent.putExtra(Constants.EXTRA_RESULT,result);
    intent.putExtra(Constants.EXTRA_A,a);
    sendBroadcast(intent); //封装数据到 Intent 中,发送 Intent 对象到广播接受者
    }catch (InterruptedException e){ e.printStackTrace(); }   }
```

步骤 6:在 MyIntentService 中新增 startMyIntentService()用于启动服务,代码如下。

```
public static void startMyIntentService(Context context, int a, int b){
   Log.i (TAG," ****************************"); Log.i (TAG,"
startMyIntentService: " + a + " + " + b);
   Intent intent = new Intent(context,MyIntentService.class);
   intent.setAction(Constants.ACTION_FOR_SUM); //封装运算参数和 action 到 Intent 对象中
   intent.putExtra(Constants.EXTRA_A,a);
   intent.putExtra(Constants.EXTRA_B,b);
   context.startService(intent); //用 startService 将 Intent 传递给 onHandleIntent()方法 } }
```

步骤 7:在 MainActivity.java 中,编写代码实现基础功能。

```
public class MainActivity extends AppCompatActivity {
    public static final String TAG = "MainActivity";private LinearLayout ll_container;
    @Override
    protected void onCreate(Bundle savedInstanceState) {
       …………
       registerBroadcast(); }
    private void registerBroadcast(){ //注册一个广播
        IntentFilter intentFilter = new IntentFilter();
        intentFilter.addAction(Constants.ACTION_RESULT);
        registerReceiver(forSumReceiver, intentFilter);Log.i(TAG,"registerBroadcast");}
```

步骤 8:在 MainActivity.java 中新增 addTask 方法代码,用于按钮“发起计算请求”的点击事件,代码如下。

```
private int a = 1;
public void addTask(View view){
    int b = new Random().nextInt(101) + 1;
    MyIntentService.startMyIntentService(this,a,b); //用户启动 IntentService 服务
    TextView textView = new TextView(this);
    textView.setTextColor(Color.RED);
    textView.setText(a + " + " + b + " = " + "  正在计算中...");
    textView.setTag(a);
    ll_container.addView(textView); //将新的运算表达式显示在界面上
    Log.i(TAG,"addTask --- " + a + " + " + b);a++;}
```

步骤 9:在 MainActivity.java 中,实现广播接受者 forSumReceiver 的功能。

```
private BroadcastReceiver forSumReceiver = new BroadcastReceiver() {
   @Override
   public void onReceive(Context context, Intent intent){
      if(intent.getAction() == Constants.ACTION_RESULT){
      int a = intent.getIntExtra(Constants.EXTRA_A,0);
      int result = intent.getIntExtra(Constants.EXTRA_RESULT,0);
      Log.i(TAG,"onReceive -- result:" + result);
      handleResult(a,result); } } };
```

步骤 10:在 MainActivity.java 中,实现 handleResult()响应广播,更新界面信息。

```
private void handleResult(int a, int result){
    TextView textView = ll_container.findViewWithTag(a);
    String old = textView.getText().toString();
    String newText = old.replaceAll(" 正在计算中...",String.valueOf(result) + "  计算完成");
    textView.setTextColor(Color.BLACK);
    textView.setText(newText);
    Log.i(TAG,"handleResult -- result:" + result);Log.i(TAG,"####################
###");}
```

步骤 11：在 MainActivity.java 中，编写 onDestroy() 方法实现线程销毁功能。

```
@Override
protected void onDestroy() { //主要功能是 Service 的销毁和注销广播接收者
    super.onDestroy();unregisterReceiver(forSumReceiver);
    Log.i(TAG,"onDestroy - - result:" + Constants.ACTION_RESULT); }}
```

运行程序，首先只点击一次按钮，界面显示和控制台输入如图 9-11 和图 9-12 所示。

图 9-11　正在计算中的界面

图 9-12　计算完成的界面

查看正在计算中和计算完成两个过程中控制台的输出信息，掌握 Service 生命周期的变化，如图 9-13 和图 9-14 所示。

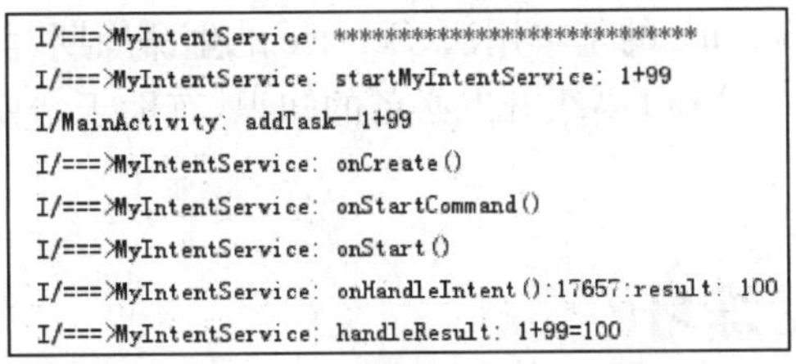

图 9-13　正在计算中的控制台输出

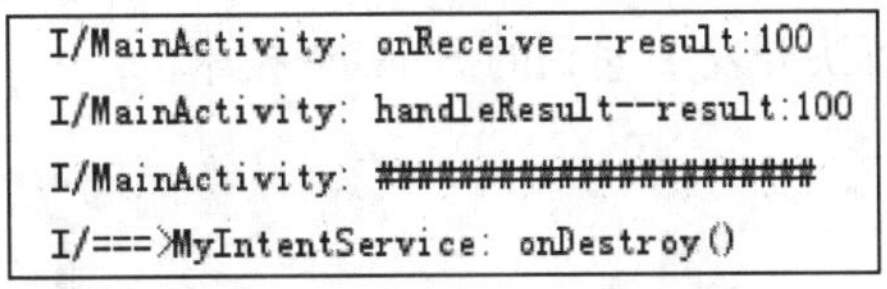

图 9-14　日志中显示的计算完成信息

继续点击发起计算请求，新的计算请求被发起，界面如图 9-15 所示。等待片刻，显示新的计算请求陆续完成，如图 9-16 所示。

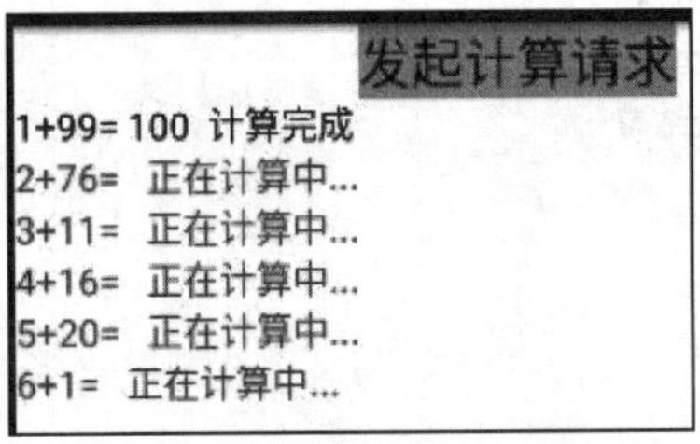

图 9-15　发起其他的计算请求

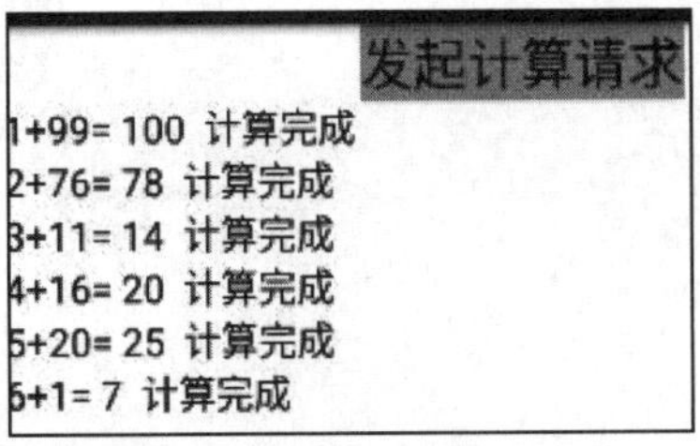

图 9-16　新发起的计算完成

```
I/===>MyIntentService: ****************************
I/===>MyIntentService: startMyIntentService: 2+76
I/MainActivity: addTask--2+76
I/===>MyIntentService: onCreate()
I/===>MyIntentService: onStartCommand()
I/===>MyIntentService: onStart()
I/===>MyIntentService: onHandleIntent():17666:result: 78
I/===>MyIntentService: handleResult: 2+76=78
I/===>MyIntentService: ****************************
I/===>MyIntentService: startMyIntentService: 3+11
I/MainActivity: addTask--3+11
I/===>MyIntentService: onStartCommand()
I/===>MyIntentService: onStart()
I/===>MyIntentService: ****************************
I/===>MyIntentService: startMyIntentService: 4+16
I/MainActivity: addTask--4+16
I/===>MyIntentService: onStartCommand()
I/===>MyIntentService: onStart()
I/===>MyIntentService: ****************************
I/===>MyIntentService: startMyIntentService: 5+20
I/MainActivity: addTask--5+20
I/===>MyIntentService: onStartCommand()
I/===>MyIntentService: onStart()
I/===>MyIntentService: ****************************
I/===>MyIntentService: startMyIntentService: 6+1
I/MainActivity: addTask--6+1
I/===>MyIntentService: onStartCommand()
I/===>MyIntentService: onStart()
```

图 9-17　日志中显示的新发起的计算请求

```
I/MainActivity: onReceive --result:78
I/MainActivity: handleResult--result:78
I/MainActivity: #####################
I/===>MyIntentService: onHandleIntent():17666:result: 14
I/===>MyIntentService: handleResult: 3+11=14
I/MainActivity: onReceive --result:14
I/MainActivity: handleResult--result:14
I/MainActivity: #####################
I/===>MyIntentService: onHandleIntent():17666:result: 20
I/===>MyIntentService: handleResult: 4+16=20
I/MainActivity: onReceive --result:20
I/MainActivity: handleResult--result:20
I/MainActivity: #####################
I/===>MyIntentService: onHandleIntent():17666:result: 25
I/===>MyIntentService: handleResult: 5+20=25
I/MainActivity: onReceive --result:25
I/MainActivity: handleResult--result:25
I/MainActivity: #####################
I/===>MyIntentService: onHandleIntent():17666:result: 7
I/===>MyIntentService: handleResult: 6+1=7
I/MainActivity: onReceive --result:7
I/MainActivity: handleResult--result:7
I/MainActivity: #####################
I/===>MyIntentService: onDestroy()
```

图 9-18　日志中显示的计算完成信息

分析以上控制台的输出信息，可以看出执行完一个 Intent 请求对象所对应的工作之后，如果没有新的 Intent 请求达到，则自动停止 Service，并且 IntentService 的所有请求如果在一个生命周期中完成，则所有请求是在一个工作线程中顺序执行的，否则是在不同的工作线程中完成。

小结

本章介绍了 Android 平台的多线程编程、Service 的基本用法、Service 的生命周期、前台 Service 和 IntentService 等。这些内容是进行 Android 高级开发必备的知识，有助于帮助读者解决开发中遇到的 Service 技术。

课后练习

【微信扫码】
第 9 章课后练习

第 10 章

网络编程技术

在当今的互联网时代,我们编写的 Android 程序基本都需要联网。网络编程是任何一个 Android 程序员都需要的技能。作为开发者,我们需要考虑如何利用网络编写更加出色的应用程序,像 QQ、微博、微信等常见的应用程序都会大量使用网络技术。本章主要介绍如何在 Android 系统中用 HTTP 协议和服务器端进行网络交互,对网络返回的数据进行解析、处理并展示在界面上。

10.1 WebView

在实际使用中,如果需要显示一个网页,可以通过 Intent 打开浏览器,交给浏览器去显示,也可以在程序中直接用 WebView 组件显示。下面通过案例来演示 WebView 控件使用。

10.1.1 WebView 的使用

【微信扫码】
案例 10 - 1
相关文件

案例 10 - 1:WebView 的使用

步骤 1:打开 activity_main.xml 编写布局文件,直接使用 WebView 组件,代码如下。

```
< WebView android: id = "@ + id /web_view" />
```

步骤 2:编写 MainActivity 文件,代码如下。

```
public class WebViewActivity extends AppCompatActivity {
    WebView webView;
    @Override
    protected void onCreate(Bundle savedInstanceState) {
        …… //调用 WebView 的 setWebViewClient()方法
        webView. setWebViewClient(new WebViewClient());
        //调用 WebView 的 loadUrl 将网址传入就可以浏览网页的内容了
        webView. loadUrl("http: //www. cslg. edu. cn"); }
```

步骤 3:修改 AndroidManifest. xml 应用配置文件,添加访问 Internet 的权限,< uses-permission android: name ="android. permission. INTERNET"/ >这个权限属于普通权限,添加后系统会自行授权。在 Android 5.0 以后默认使用 HTTPS 方式,为允许程序使用 HTTP 方式访问 Internet,在 < Application ></ Application > 中添加 android: usesCleartextTraffic ="true"。

这样编写完成之后，WebView 就可以使用，运行程序，显示效果如图 10－1 所示。可以看到，程序已经具备了一定的简单网络浏览功能。

但是如果仔细看的话，页面中有些部分并没有显示出来。这不是网络不通畅造成的，而是因为网页上有很多 JavaScript 程序和其他的一些插件等，WebView 还需要进行相应的配置，才能更好地显示。下面就对 WebView 进行配置。

步骤 4：在 MainActivity 中添加方法 webViewSetting()，里面对 WebView 进行配置。核心代码如下。

```
private void webViewSetting() {
    WebSettings webSettings = webView.getSettings();
    …………
    webSettings.setJavaScriptEnabled(true); //是否支持 JS
    webSettings.supportMultipleWindows(); //支持多窗口
    webSettings.setJavaScriptCanOpenWindowsAutomatically(true); //允许 JS 打开新窗口
    webSettings.setLoadsImagesAutomatically(true); //支持自动加载图片
    String cacheDirPath = getCacheDir().getAbsolutePath();}
```

在 onCreate()方法中，在 loadUrl()方法之前添加代码 webViewSetting()用于启动配置。再次运行程序如图 10－2 所示，可以看到，使用 JavaScript 插件的网页部分也能流畅显示了。

图 10－1　WebView 加载网页

图 10－2　配置 WebView 的网页浏览

10.1.2　WebViewClient

步骤 5：WebViewClient 可以帮助 WebView 处理各种通知和请求事务、核心代码如下。

```
WebViewClient mWebViewClient = new WebViewClient() {
    public boolean shouldOverrideUrlLoading(WebView view,String url){
      view.loadUrl(url);return true;} //获取 url,查看 url.contains("add"),进行添加操作
    @Override
     public boolean shouldOverrideUrlLoading (WebView view, WebResourceRequest request)
{view.loadUrl(request.getUrl().toString()); return true; }
    @Override
    public void onReceivedSslError(WebView view, SslErrorHandler handler, SslError error)
{ //重写此方法可以让 webview 处理 https 请求.
        super.onReceivedSslError(view, handler, error);  }
    @Override
     public void onScaleChanged ( WebView view, float oldScale, float newScale) { super.
onScaleChanged(view, oldScale, newScale); } //WebView 发生改变时调用
    …… //省略部分方法 };
```

如果修改为 webView.setWebViewClient(new WebViewClient()) 或者 webView.setWebViewClient(mWebViewClient)可以启动配置参数。

10.1.3　WebChromeClient

步骤 6:WebChromeClient 是辅助 WebView 处理 JavaScript 的对话框、网站图标、网站 title、加载进度等,代码如下。

```
WebChromeClient mWebChromeClient = new WebChromeClient() {
    @Override
    public void onProgressChanged(WebView view,int newProgress){ } //网页加载进度
    @Override
    public void onReceivedTitle(WebView view, String title) { } //获取 Web 页中的 title 用来
设置自己界面中的 title
    @Override
    public void onReceivedIcon(WebView view, Bitmap icon){     } //获取 Web 页中的 icon
    @Override
    public boolean onJsAlert(WebView view, String url, String message, JsResult result) {
return true; } //处理 alert 弹出框
    @Override
     public boolean onJsPrompt (WebView view, String url, String message, String defaultValue,
JsPromptResult result) {return true;} //处理 confirm 弹出框
    @Override
    public boolean onJsConfirm(WebView view, String url, String message, JsResult result) {
return true;} //处理 prompt 弹出框 };
```

webView.setWebChromeClient(mWebChromeClient)用于启动 WebChromeClient 方法。

WebView 还有很多更加高级的使用技巧,我们就不再继续进行探讨了。真正的网络开发,还需要了解 HTTP 协议等知识,下面就对 HTTP 协议进行介绍。

10.2 HTTP

上一节中使用到的 WebView 控件,其实就是我们向服务器发起了一条 HTTP 请求,接着服务器分析出我们想要访问的是 CSLG 的首页,于是会把该网页的 HTML 代码进行返回,然后 WebView 再调用手机浏览器的内核,对返回的 HTML 代码进行解析,最终将页面展示出来。简单来说,WebView 已经在后台帮我们处理好了发送 HTTP 请求、接收服务响应、解析返回数据以及最终的页面展示这几步工作,接下来就让我们深入地理解一下 HTTP 协议到底是什么。

10.2.1 HTTP 协议

HTTP 的全称是 Hypertext Transfer Protocol(超文本传输协议)。HTTP 协议规定客户端和服务器之间的数据传输格式,让客户端和服务器能有效地进行数据沟通。HTTP 报文由从客户机到服务器的请求和从服务器到客户机的响应构成。请求报文格式如图 10-3 所示。

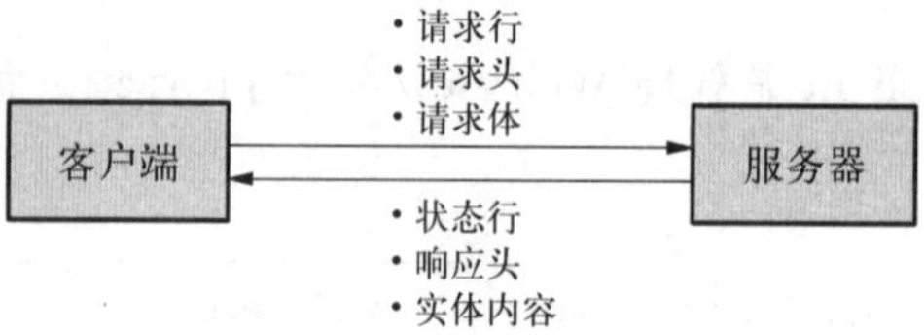

图 10-3 HTTP 通信过程

HTTP 是基于客户/服务器模式,且面向连接的。客户与服务器之间的 HTTP 连接是一次性连接,它限制每次连接只处理一个请求,当服务器返回本次请求的应答后便立即关闭连接,下次请求再重新建立连接。现在用的 HTTP 协议的版本号是 1.1。在HTTP 1.1协议中,定义了九种发送 HTTP 请求的方法,分别为 GET, POST, OPTIONS, HEAD, PUT, DELETE, TRACE, CONNECT, PATCH。根据 HTTP 协议的设计初衷,不同的方法对资源有不同的操作方式,最常用的是 GET 和 POST。

10.2.2 GET 和 POST

GET:在请求 URL 后面加"?"然后列出发给服务器的参数,多个参数之间用"&"隔开,如 http://www.cslg.com/login? username=123&pwd=234&type=JSON。

POST:发给服务器的参数全部放在请求体中。理论上 POST 传递的数据量没有限制(具体还得看服务器的处理能力)。如果增加、修改、删除数据,建议使用 POST。如果要传递大量数据,比如文件上传,只能用 POST 请求。包含机密或者敏感信息的数据,建议用 POST。

下面我们来学习如何提交 HTTP GET 和 HTTP POST 的请求来获得 HTTP 资源协议。

案例 10－2：使用 HTTP GET 和 HTTP POST 请求获得 HTTP 资源

【微信扫码】
案例 10－2
相关文件

首先我们需要有一个 HTTP 资源，为了方便使用，我们自己构建一个 HTTP 站点。使用 JSP/Servlet 技术开发，Apace Tomcat 提供网站服务。

步骤 1：建立 HTTP 资源站点。在 J2EE Eclipse 中，点击 File→New→Dynamic Web Project，新建一个 Web 站点，命名为 WebHttpGetPost。在项目文件夹上，右键单击 New/Servlet，新建一个 Servlet 程序，代码如下。

```
public class QueryServlet extends HttpServlet{
    private static final long serialVersionUID = 1L;PrintWriter out = null;
    @Override
    protected void service(HttpServletRequest request,HttpServletResponse response) throws
ServletException, IOException{
        response.setContentType("text/html;charset = utf - 8"); String queryStr = "";
        if ("post".equals(request.getMethod().toLowerCase()))
        queryStr = "获得 POST 请求;网页传递的字符串是" + new String(request.getParameter("
webstr").getBytes("iso - 8859 - 1"), "utf - 8");
        else if ("get".equals(request.getMethod().toLowerCase()))
        queryStr = "GET 请求;网页传递的字符串是" + request.getParameter("webstr");
        out = response.getWriter();out.println(queryStr);  }}
```

代码比较简单，在方法 service()中，根据不同的传递请求，返回不同的响应字符串。

步骤 2：配置 Web 的 Servlet，在 WebContent/WEB-INF 文件夹中修改 web.xml 文件。

```
<servlet><display-name>YJYServlet</display-name><servlet-name>YJYServlet</
servlet-name><servlet-class>com.yangjianyong.mobile.QueryServlet</servlet-class></
servlet>
 <servlet-mApping><servlet-name>YJYServlet</servlet-name><url-pattern>/YJYServlet</
url-pattern></servlet-mApping>
```

这里主要是定义了 Servlet 的映射关系。相关内容请学习 JSP/Servlet 相关知识。

通过步骤 1 和 2，我们建立了一个 HTTP 的资源站点。通过接收外部的 Servlet 请求来提供服务。我们还可以新建一个 JSP 页面来测试。Query.jsp 页面代码请参考本书提供的资源。

步骤 3：建立 Web 站点。

编辑好的 JSP Servlet 项目，需要导出并放入 Apache 服务器中才能使用。先选中 WebHttpGetPost 项目文件夹，点击 File→Export→Web→WAR file，导出为 WAR 格式。

在 D 盘中新建一个文件夹，命名为 AndroidStudioWeb。将 WAR 文件放入后解压缩。

在 Apache Tomcat 安装目录里，选择 conf→server.xml，在 Host 标签内，添加代码如下。

```
<Context path = "/getpost"  docBase = "D:\AndroidStudioWeb\YJYWebHttpGetPost" />
```

Context path 为访问的上下文，docBase 为文件所在的位置。配置好之后，重启 Apache 服务器，在浏览器中输入 http://localhsot:8080/getpost/query.jsp 能查看站点，如图 10 - 4 所示。

图 10 - 4　Web 站点的显示

接下来设计安卓端的程序，来连接 HTTP 站点获得服务资源。在 Android Studio 中新建一个项目，命名为 Web_Teach_GetPost。

步骤 4：编写布局文件，放置文本框和两个按钮，分别用来提交 Get 和 Post 请求。

```
<Button android:id = "@ + id /btnGetQuery"   android:text = "GET 提交"  />
<Button android:id = "@ + id /btnPostQuery"   android:text = "POST 提交"  />
<TextView android:id = "@ + id /tvQueryResult"    android:text = "显示的信息" />
```

步骤 5：修改 MainActivity 代码。

```
public class MainActivity extends Activity implements View. OnClickListener {
    String url = "http: //10. 18. 45. 87:8080 /getpost /YJYServlet"; //访问 HTTP 资源的 URL,
URL 必须使用 IP 地址,不能用 localhost 代替
    HttpResponse httpResponse = null;TextView tvQueryResult;EditText etBookName;
    @Override
    public void onCreate(Bundle savedInstanceState){
            …… btnGetQuery. setOnClickListener ( this ); btnPostQuery. setOnClickListener
(this);}
```

步骤 6：新建一个方法 onClick(View view)实现两个按钮事件，先实现 Get 提交方式。

```
switch (view. getId()){
  case R. id. btnGetQuery:new Thread(new Runnable() {
     @Override
    public void run() {
     url + = "?webstr = " + etBookName. getText (). toString(); //拼接文本框输入的内容,构造 URL
     HttpGet httpGet = new HttpGet(url); //将要请求的 URL 通过构造方法传入 HttpGet 对象
    try {httpResponse = new DefaultHttpClient(). execute(httpGet); 使用 DefaultHttpClient
类的 execute 方法发送 http Get 请求,并返回 httpResponse 对象
     } catch (IOException e) {e. printStackTrace();}
    runOnUiThread(new Runnable() { //使用 runOnUiThread 子线程更新 UI
      @Override
      public void run() {
        String result = null; //判断返回的 httpStatus,SC_OK 就执行更新
```

```
        if (httpResponse.getStatusLine().getStatusCode() == HttpStatus.SC_OK){
            try { result = EntityUtils.toString(httpResponse.getEntity());
                tvQueryResult.setText(result.replaceAll("\r", ""));
            } catch (IOException e) {e.printStackTrace(); } }}});}}).start();break;
```

步骤 7:完成 Post 提交方式代码。

```
case R.id.btnPostQuery:new Thread(new Runnable() {
    @Override
    public void run() {   try{ HttpPost httpPost = new HttpPost(url);
     List< NameValuePair > params = new ArrayList< NameValuePair >();
     params.add(new BasicNameValuePair("webstr", etBookName.getText().toString()));
     httpPost.setEntity(new UrlEncodedFormEntity(params, HTTP.UTF_8));
     httpResponse = new DefaultHttpClient().execute(httpPost);
     runOnUiThread(new Runnable() {
  @Override
  public void run() {
     String result = null;
       if (httpResponse.getStatusLine().getStatusCode() = = 200){
         try {result = EntityUtils.toString(httpResponse.getEntity());
          } catch (IOException e) {e.printStackTrace();}
         tvQueryResult.setText(result.replaceAll("\r", ""));} }});
  }  catch (IOException e) {e.printStackTrace();}  }}).start(); break;
```

步骤 8:配置项目信息。在 AndroidManifest.xml 中添加如下代码。

(1) 网络权限:< uses-permission android:name="android.permission.INTERNET" / >。

(2) 允许应用程序使用 HTTP 方式访问网络:< android:usesCleartextTraffic ="true">。

(3) 对 Apache 的支持:< uses-library android: name =" org. apache. http. legacy" android:required ="false"/ >。同时在 build.grade(Module: App)的 android {}中添加对 Apache 自定义库的支持代码 useLibrary 'org.apache.http.legacy'。

运行程序,效果显示如图 10-5 和图 10-6 所示。

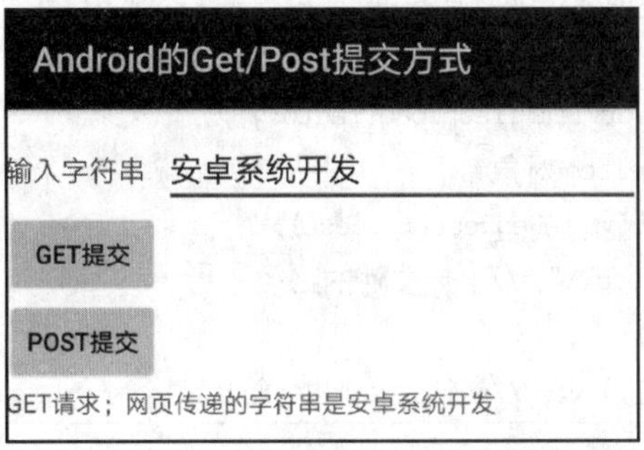

图 10-5　Get 提交方式

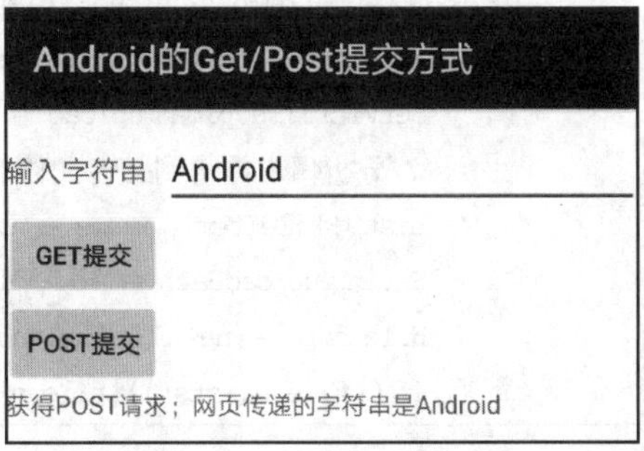

图 10-6　Post 提交方式效果

10.3 HttpURLConnection

Android 系统发送 HTTP 请求一般有两种方式：HttpURLConnection 和 HttpClient。目前 HttpClient 已被完全移除，官方建议使用 HttpURLConnection，下面学习它的使用。

10.3.1 上传资源到服务器

HttpURLConnection 是一种多用途、轻量级的 HTTP 客户端，使用它来进行 HTTP 操作适用于大多数的应用程序。使用 HttpURLConnection 访问 HTTP 资源，包含如下几步：① 使用 java.net.URL 封装 HTTP 资源的 URL，并使用 openConnection 方法获得 HttpURLConnection 对象。② 设置请求方法，例如 GET、POST 等。③ 设置输入/输出及其他权限，比如需要下载 HTTP 资源或向服务端上传数据等。

下面我们通过一个案例来学习 HttpURLConnection 的使用。

案例 10－3：使用 HttpURLConnection 进行网络传输

本案例使用 HttpURLConnection 实现了一个上传文件的应用程序。该程序可以将手机上的文件上传到服务端，需要分别实现服务器端 Web 系统和移动端的安卓程序。

【微信扫码】
案例 10－3
相关文件

步骤 1：在 Eclipse 中点击 File→New→Dynamic Web Project 新建一个 Web 站点，命名为 WebUpload。在项目文件夹上右键单击 New→Servlet，新建一个 Servlet 程序。

```
public class UploadServlet extends HttpServlet{
    private static final long serialVersionUID = 1L;
     protected void service (HttpServletRequest request, HttpServletResponse response)
throws ServletException, IOException{
   try{request.setCharacterEncoding("UTF-8"); //设置处理请求参数的编码格式
       response.setContentType("text/html;charset=UTF-8");
       PrintWriter out = response.getWriter();
           //使用 Commons-UploadFile 组件处理上传的文件数据
           FileItemFactory factory=new DiskFileItemFactory(); //FileItemFactory
           ServletFileUpload upload = new ServletFileUpload(factory);
           //分析请求并得到上传文件的 FileItem 对象
           List<FileItem> items = upload.parseRequest(request);
           String uploadPath = "d:\\YJYUpload\\"; //上传文件的路径
           File file = new File(uploadPath);
           if (!file.exists()){file.mkdir();}…… //文件上传功能(略)   }
```

步骤 2：服务器端的 Web.xml 文件。

```
< servlet >< servlet-name > UploadServlet </ servlet-name >< servlet-class > com.yangjianyong.mobile.UploadServlet </servlet-class ></servlet >
< servlet-mApping >< servlet-name > UploadServlet </ servlet-name >< url-pattern >/YJYUploadServlet </url-pattern ></servlet-mApping >
```

编写好之后,发布在 Apache Tomcat 服务器上。

下面实现移动端的安卓程序。首先浏览 SD 卡中的文件,当单击某个文件后,程序能自动连接服务器并将文件上传。在 Android Studio 中新建一个项目,命名为 Web_Teach_FileUpload。

步骤 3:编写 SD 卡的文件浏览功能和权限代码,在 src 中新建一个文件夹 widget,里面包含一个接口 OnFileBrowserListener 和类 FileBrowser,在 app/java 下创建权限的控制类 MPermissionsActivity,请参阅第 7 章案例 PM_Teach_PermissionDemo。

步骤 4:编写布局文件。

```
< android. support. constraint. ConstraintLayout    xmlns: mobile = " http: // filebrowser.yangjianyong.com" …… >< LinearLayout    android:orientation = "vertical">
    < com. yangjianyong. filebrowser. widget. FileBrowser android: id = "@ + id /filebrowser1" mobile: folderImage = "@drawable /folder"    mobile: extName1 = " jpg"    mobile: fileImage1 = "@drawable /jpg"    mobile:otherFileImage = "@drawable /other"    />
< /LinearLayout >< /android. support. constraint. ConstraintLayout >
```

步骤 5:修改 MainActivity 代码。

```
public class MainActivity extends MPermissionsActivity implements OnFileBrowserListener {
    private final int SDK_PERMISSION = 1; //权限获取的返回值
    String uploadUrl = "http: //10.18.45.87:8080 /upload /YJYUploadServlet";
    String end = "\r\n";String twoHyphens = "--"; //两个字符的连接符
    String boundary = " ****** "; //字符串分界符
    @Override
    protected void onCreate(Bundle savedInstanceState){
        …………
      if (super. checkSelfPermission ( " android. permission. WRITE_ EXTERNAL_ STORAGE") ! = PackageManager. PERMISSION_GRANTED) {
      getPermissions(); //如果读写权限没有开启则通过 getPermissions()获得权限
      } else { //实现手机 SD 卡文件目录浏览,设置事件监听
      FileBrowser fileBrowser1 = findViewById(R. id. filebrowser1);
      fileBrowser1. setOnFileBrowserListener(this); } }
```

步骤 6:功能实现——权限申请和获取,在 MainActivity 中,添加 getPermissions()和 onRequestPermissionsResult()两个方法,代码请参照 7.3.3 对权限进行轻量级封装。

步骤 7:功能实现——点击文件进行上传。MainActivity 类中继承了 OnFileBrowserListener,这里完善其方法 onFileItemClick()。

```
    @Override
    public void onFileItemClick(final String filename){
        setTitle(filename);
        new Thread(new Runnable() { //启动线程完成网络访问和 UI 更新
            @Override
            public void run() {
                try{……}
    //实现点击某个文件,将文件名显示在标题栏中
    public void onDirItemClick(String path){ setTitle(path);   }
```

运行程序,首先启动 Apache Tomcat 服务器,然后运行安卓程序,打开手机的 SD 卡目录,如图 10-7 所示,点击某个文件,文件会自动上传到服务器上,打开服务器上的 D 盘,在目录 YJYUpload 中能找到上传的文件,如图 10-8 所示。

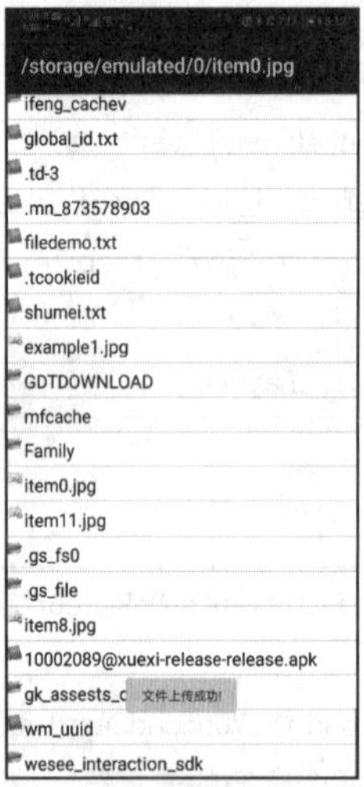

图 10-7　打开手机浏览 SD 目录

图 10-8　上传文件效果

10.3.2　从服务器获取资源

一个软件拥有网络功能的时候,经常需要从网络上获得数据,并在控件中显示网络数据。下面我们通过一个案例来学习如何从网络上获得资源并显示这些资源数据。

案例 10-4:网络数据的获取和加载

本案例实现给应用程序打分。在网络服务器上提供这些应用程序的名字、星级等信息,安卓应用程序从网络上获得这些信息,并将其显示在界面上。

【微信扫码】案例 10-4 相关文件

步骤 1:配置服务器上的资源信息。

在 Apache Tomcat 服务器的 Web 目录中,新建一个虚拟目录,命名为 WebImageList,包含 META-INF、WEB-INF 两个文件夹,文本文件 list.txt 和若干个图标,如图 10-9 所示。

图 10－9　WebImageList 项目的 Web 目录

list.txt:该文件指定了应用程序的图标、名称和评价分数,中间用逗号隔开。

```
calendar.png,我的日历,1    earth.png,地球仪,2    game.png,在线游戏,4    qq.png,QQ,4.5
jscb.png,词霸,5    cube.png,魅力魔方,3    ddz.png,斗地主,2.5    java.png,Java 语言,3.5
```

下面实现安卓客户端的功能,在 Android Studio 中新建一个项目,命名为 Web_Teach_ImageList。

步骤 2:在 app/res/layout/activity_main.xml 中编写布局文件,放置 ImageView 和 TextView 用来显示图片和文字,靠右侧排列一组 TextView 和 RatingBar 用来显示数字和图形。

```
<ImageView android:id="@+id/ivLogo" />
<RelativeLayout  android:orientation="vertical"  android:gravity="right" >
        <TextView android:id="@+id/tvApplicationName" />
</RelativeLayout>
<RelativeLayout  android:orientation="vertical"  >
      <TextView android:id="@+id/tvRating"  android:text="5.0" />
      <RatingBar android:id="@+id/ratingbar"  android:numStars="5" />
</RelativeLayout>
```

步骤 3:在 src 中新建一个名为 NetUtils 工具类,里面封装 HttpURLConnection 的方法。

```
public class NetUtils {
   public static InputStream getInputStream(String url){
     HttpURLConnection conn = null;
     try {
       URL mURL = new URL(url); //创建一个 URL 对象
       conn = (HttpURLConnection)mURL.openConnection(); //获取 HttpURLConnection 对象
     …………return is;
     }else{throw new NetworkErrorException("response status is " + responseCode); }
     } catch (Exception e) {  e.printStackTrace();
     } finally {if (conn != null){conn.disconnect(); //关闭连接 } }
       return null; }
    public static Bitmap getHttpBmp(String url) { //转换为 Bitmap 用来显示图片
        Bitmap bm = null;InputStream is = getInputStream(url);
        bm = BitmapFactory.decodeStream(is);return bm; }}
```

步骤 4:编写数据处理适配器类。

在 src/java 中,新建类 ImageListAdapter。本例的核心是负责处理数据的 ApkListAdapter 类,该类是 BaseAdapter 的子类。

```
public class ImageListAdapter extends BaseAdapter   {
    private String inflater = Context.LAYOUT_INFLATER_SERVICE;
    private String rootUrl = "http://10.18.45.87:8080/imagelist/";
     private List < ImageListAdapter.ImageData > imageDataList = new ArrayList <
ImageListAdapter.ImageData >();
    …… //声明其他控件和组件
    public void initView(){…… //初始化控件}
    class ImageData{ //ImageData 内嵌类,用于保存图像文件名、应用程序名和评价分数
        public String url;public String ApplicationName;public float rating; }
    public ImageListAdapter(Context context){ //类的构造方法
        this.context = context;
        layoutInflater = (LayoutInflater)context.getSystemService(inflater);
         //获得了 list.txt 文件的内容,根据分隔符","按行读取内容并拆分后将其保存在集合
List< lmageData >对象中
        new Thread(new Runnable() {   ……   }).start();}
    …… //补充 BaseAdapter 的其他三个抽象方法 getCount()、getItem()、getItemId()
    @Override
    //根据 List< ImageData >对象中的图像信息下载相应的图像文件,并返回显示这些图像的
ImageView 对象
    public View getView(int position, View convertView, ViewGroup parent){
        ……
        imagehandler = new Handler(); //启动新的线程下载图标
        new Thread(new Runnable() {
            @Override
            public void run() {
                bm = new NetUtils().getHttpBmp(bitmapurl);
                imagehandler.post(new Runnable() {
                    @Override
                    public void run() {ivLogo.setImageBitmap(bm);} }); } }).start();
        return linearLayout; }}
```

步骤 5:修改 MainActivity 代码。

```
public class MainActivity extends ListActivity {
    @Override
    public void onCreate(Bundle savedInstanceState){
        super.onCreate(savedInstanceState);
        ImageListAdapter imageListAdapter= new ImageListAdapter(MainActivity.this);
        setListAdapter(imageListAdapter);   }}
```

代码中直接调用适配器类，实现网络资源调用和控件数据填充，运行程序，效果如图 10－11 所示。从运行的界面可以看到，应用程序的文字和评价星级都可以正常显示出来。但是应用程序对应的图标却没显示出来，查看 Android Studio 的 Run，显示如图 10－10 所示错误信息。

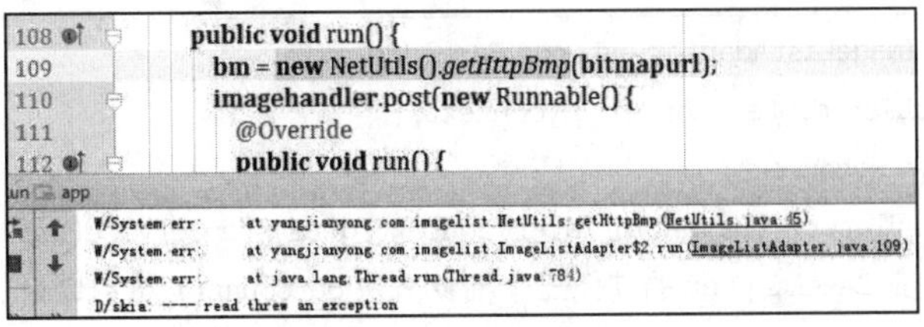

图 10－10　Web_Teach_Image List 的错误信息

分析原因，我们得知程序启动子线程，根据 URL 地址和文本信息，可以获得网络资源信息，应用程序的名称和评价星级。但是获得程序的图标还要再次连接网络，相当于在子线程中再次启动子线程，这样使用原先的做法是实现不了的。那么遇到这样的场景，怎么解决呢？下面介绍开源项目 SmartImageView。

10.3.3　SmartImageView 和 AsyncHttpClient

SmartlmageView 主要是为了加速从网络上加载图片，它继承自 Android 自带的 ImageView 组件，支持加载通信录中的图片，支持异步加载图片，支持图片缓存等。

每次加载图片都需要开启子线程，这样是非常麻烦的，为此可以使用开源项目 AsyncHttpClient。AsyncHttpClient 可以处理异步 HTTP 请求，并通过匿名内部类处理回调结果，HTTP 异步请求均位于非 UI 线程中，不会阻塞 UI 操作，AsyncHttpClient 通过线程池处理并发请求，处理文件上传、下载，响应结果自动打包成 JSON 格式，使用起来非常方便。下面使用 AsyncHttpClient＋ SmartlmageView 来实现网络图片资源的加载问题。

案例 10－5：SmartlmageView 和 AsyncHttpClient 的使用

新建一个 Android 项目，命名为 Web_Teach_SmartlmageView。

步骤 1：导入项目所需要的 jar 包，SmartlmageView 控件的 jar 包的下载地址为 http://loopj.com/android-smart-image-view/。在 build.gradle(Module：App) 中添加项目依赖，导入 AsyncHttpClient 和 SmartlmageView 的 jar 包。

【微信扫码】
案例 10－5
相关文件

```
implementation files('libs /android-async-http-1.4.8.jar')
implementation files('libs /android-smart-image-view-1.0.0.jar') }
```

步骤 2：打开 app/res/layout/activity_main.xml 编写布局文件，同上一个案例的布局文件相似，将< ImageView/ >替换为下面的代码，注意引用 SmartImageView 需要写出完整路径。

```
< com.loopj.android.image.SmartImageView   android:id = "@ + id /ivLogo" ……  />
```

步骤 3：创建 ImageList 的实体类。

```
public class ImageData {
    public String url; public String ApplicationName; public float rating; }
```

步骤4:修改MainActivity.java,功能实现——定义变量(略)。

步骤5:在MainActivity.java中,新增方法initView用于初始化控件功能实现(略)。

步骤6:在MainActivity.java中新增一个继承BaseAapter类ImageListAdapter。

```
private class ImageListAdapter extends BaseAdapter{
  public ImageListAdapter(Context mycontext){
      context = mycontext;
      layoutInflater = (LayoutInflater)context.getSystemService(inflater); }
    …… //补充BaseAdapter的其他三个抽象方法getCount()、getItem()、getItemId()
    @Override
    public View getView(int position, View convertView, ViewGroup parent){
      linearLayout = (LinearLayout)layoutInflater.inflate(R.layout.activity_main, null);
      initView();
      //SmartImageView加载图片,第一个参数为图片地址,第二个参数为加载失败时显示的图片,第三个参数为加载中显示的图片
    ivLogo.setImageUrl(rootUrl + imageDataList.get(position).url, R.drawable.ic_launcher_background, R.drawable.ic_launcher_background);
    …… //加载其他控件的数据
    return linearLayout;   } }}
```

步骤7:功能实现——异步处理网络资源。

在MainActivity.java中,新增一个内部方法fillData(),代码如下。

```
private void fillData() {
    AsyncHttpClient client = new AsyncHttpClient();
    client.get(listUrl, new AsyncHttpResponseHandler() { //创建AsyncHttpClient实例
    @Override
    public void onSuccess(int i, Header[] headers, byte[] bytes) {
        try { //请求成功时加载网络资源数据
        String isr = new String(bytes, "GBK"); String s = "";
        String[] sdata = isr.split("\\r\\n");
        for (int index = 0; index < sdata.length; index++){
           String[] data = sdata[index].split(",");
           if (data.length > 2){
            ImageData imageData = new ImageData(); imageData.url = data[0]; imageData.ApplicationName = data[1]; imageData.rating = Float.parseFloat(data[2]); imageDataList.add(imageData); } }
        ImageListAdapter imageListAdapter = new ImageListAdapter(MainActivity.this);
        setListAdapter(imageListAdapter);
        }catch(UnsupportedEncodingException e){e.printStackTrace();} }
      @Override
      public void onFailure(int i, Header[] headers, byte[] bytes, Throwable throwable) { //请求失败 } });  }
```

步骤 8:功能实现——onCreate()方法。

```
@Override
protected void onCreate(Bundle savedInstanceState) {
    super.onCreate(savedInstanceState);fillData(); }
```

运行程序,图片可以显示出来,达到了预期的目的,显示如图 10－12 所示。

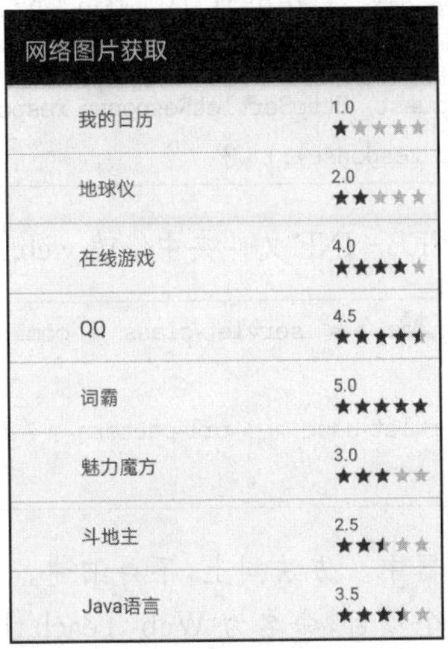

图 10－11　网络资源获取的运行效果图

图 10－12　网络图片资源的加载

10.4　JSON

Android 交互数据主要有两种方式:JSON 和 XML。JSON 是一种轻量级的数据格式。目前绝大多数公司都使用 JSON 进行数据交互。

JSON 主要通过“{}”和“[]”包裹数据,“{}”里面存放 key-value 键值对,“[]”里存放数组。标准 JSON 格式中 key 必须用双引号。要想从 JSON 中获得数据,需要对 JSON 进行解析。解析 JSON 数据可以使用 JSONObject,也可以使用开源库 GSON。下面介绍 JSONObject。

10.4.1　使用 JSONObject 解析 JSON 数据

解析 JSON 数据的原理比较简单,遇到[]时用 JSONArray,遇到{}时用 JSONObject。

案例 10－6:使用 JSONObject 解析 JSON 格式的数据

首先构建一个 HTTP 站点。使用 JSP/Servlet 技术开发,Apace Tomcat 提供网站服务。

步骤 1:建立 HTTP 资源站点。在 J2EE Eclipse 中,点击 File→New→Dynamic Web Project,新建一个 Web 站点,命名为 WebJSON。在项目文件夹上,右键单击 New/Servlet,新建一个 Servlet 程序,将 JSON 写入 Java 程序中,

【微信扫码】
案例 10－6
相关文件

使用 out.print 输出,代码如下。

```
public class Json extends HttpServlet {
    private static final long serialVersionUID = 1L; public Json() {super();}
    protected void doGet(HttpServletRequest request, HttpServletResponse response) throws
ServletException, IOException {
        ServletOutputStream out = response.getOutputStream();
        out.print("[{\"city\":\"beijing\",\"code\":\"010\"},{\"city\":\"shanghai\",
\"code\":\"021\"},{\"city\":\"xian\",\"code\":\"029\"}]");    }
    protected void doPost(HttpServletRequest request, HttpServletResponse response) throws
ServletException, IOException {   doGet(request, response);}   }
```

步骤 2:配置 Web 的 Servlet。在 WebContent/WEB-INF 文件夹中修改 web.xml 文件。

```
<servlet><servlet-name>Json</servlet-name><servlet-class>com.yjyandroid.
yjywebjson.Json</servlet-class></servlet>
<servlet-mApping><servlet-name>Json</servlet-name><url-pattern>/Json</url-
pattern></servlet-mApping>
```

编辑好的 JSP Servlet 项目,放入 Apache 服务器中。方法同上,不再细述。
接着设计安卓端,在 Android Studio 中新建一个项目,命名为 Web_Teach_JSON。
步骤 3:编写布局文件,放置一个 TextView 用来显示信息。
步骤 4:修改 MainActivity.java 代码。

```
public class MainActivity extends Activity {
  HttpResponse response;HttpGet request;TextView text;String s;
  @Override
  protected void onCreate(Bundle savedInstanceState) {
    …………
    new Thread() {
      public void run() { try {
            request = new HttpGet("http://10.18.45.87:8080/webjson/Json");
            response = new DefaultHttpClient().execute(request);
          if (response.getStatusLine().getStatusCode() = = 200) {
              String msg = EntityUtils.toString(response.getEntity());
              JSONArray array = new JSONArray(msg); s = "";
              for (int i = 0; i < array.length(); i++) {
                JSONObject o = (JSONObject) array.get(i);
                s = s + o.getString("city") + ":  " + o.getString("code") + "\n";}
              runOnUiThread(new Runnable() {
                @Override
                public void run() {text.setText(s);} }); }
            } catch (Exception e) {e.printStackTrace();}} }.start(); }}
```

这里是将服务器返回的数据传入 JSONArray 对象中。然后循环遍历这个 JSONArray,从中取出的每一个元素都是一个 JSONObject 对象,调用 getString() 方法将这些数据取出,并打印出来即可。运行程序,结果如图 10－13 所示。

图 10－13　使用 JSONObject 解析 JSON 数据

10.4.2　使用 GSON 解析 JSON 数据

JSONObject 解析 JSON 数据已经非常简单了,但 Google 提供的 GSON 开源库可以让解析 JSON 数据的工作更加简单。GSON 库可以将一段 JSON 格式的字符串自动映射成一个对象,从而不需要再手动编写代码进行解析了。比如一段 JSON 格式的数据如下:{"name":"Tom","age":20},那就可以定义一个 Person 类,并加入 name 和 age 这两个字段,然后只需简单地调用如下代码就可以将 JSON 数据自动解析成一个 Person 对象了。

```
Gson gson = new Gson(); Person person = gson.fromJson(jsonData,Person.class);
```

如果需要解析的是一段 JSON 数组,需要借助 TypeToken 将期望解析成的数据类型传入 fromJson()方法中,代码如下。

```
List<Person> people = gson.fromJson(jsonData, new TypeToken<List<Person>>(){}.
getType()};
```

下面通过一个案例演示真实的网络请求,并在案例中使用 GSON 来处理 JSON 数据。

网上有很多提供免费数据的 API,它们以 JSON 或者 XML 格式提供数据,这里使用聚合数据,网址是:https://www.juhe.cn/。在主页上点击 API,可以看到所提供的数据界面,如图 10－14 所示。

调用第一排中的第三个手机号码归属地的数据。在使用数据之前,必须注册聚合账号,填写完整的个人信息,如图 10－15 所示,这样就可以申请调用数据。

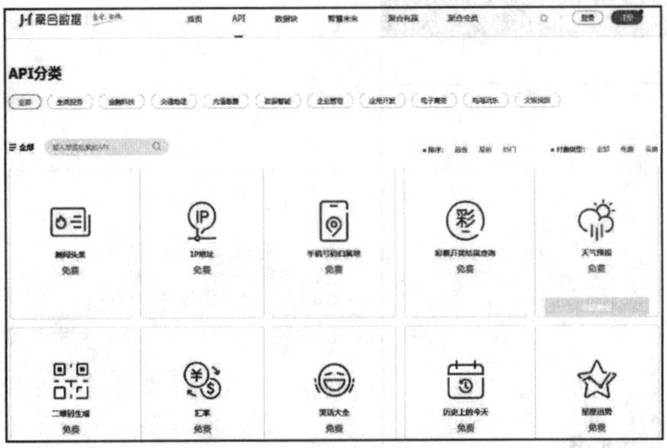

图 10－14　聚合数据的 API 分类界面

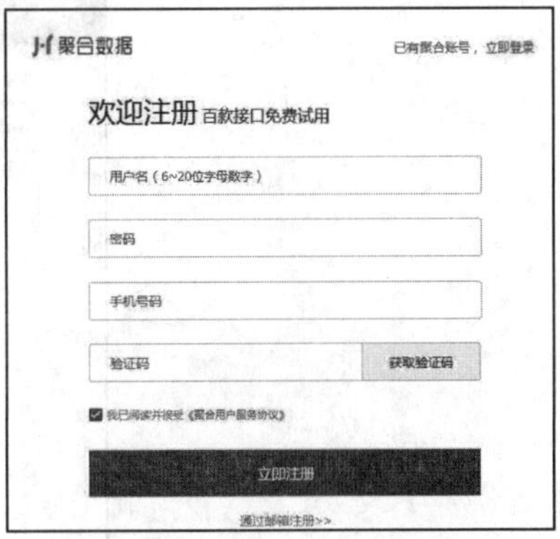

图 10－15　聚合数据的注册界面

注册完成后,进入个人中心,在"数据中心"界面申请手机号码归属地,成功之后在"我的接口”查看，可以看到账号的 AppKey，申请好之后，就可以进行开发工作了。

案例 10－7：使用 GSON 解析 JSON 格式的数据

在 Android Studio 中新建一个项目，命名为 Web_Teach_Juhe_mobile。

使用聚合数据提供的手机号码归属地 API，实现在 Android 系统中根据手机号码(段)，查询手机号码归属地信息，如省份、城市、运营商等功能。

【微信扫码】案例 10－7 相关文件

步骤 1：API 数据分析。开发之前，必须先搞清楚数据是怎么请求的。在数据界面中点击 API 文档查看，如图 10－16 和图 10－17 所示。

图 10－16　API 详情

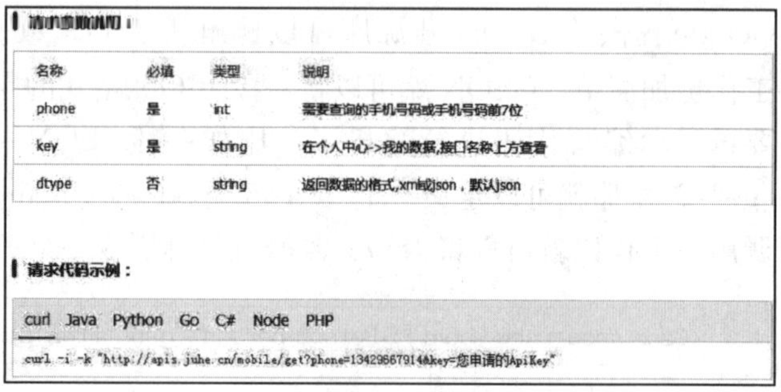

请求参数说明：
名称 必填 类型 说明
phone 是 int 需要查询的手机号码或手机号码前7位
key 是 string 在个人中心->我的数据,接口名称上方查看
dtype 否 string 返回数据的格式,xml或json，默认json
请求代码示例：
curl Java Python Go C# Node PHP
curl -i -k "http://apis.juhe.cn/mobile/get?phone=13429667914&key=您申请的ApiKey"

图 10－17　请求代码说明

页面上介绍得很清楚，使用 http://apis.juhe.cn/mobile/get? phone＝13429667914&key＝申请的 Key 发送请求。

步骤 2：导入 jar 包和添加项目依赖，项目中需要使用 GSON 库，可以去 https://github.com/google/gson 下载。下载之后，将 libs 中 GSON 的 jar 包复制粘贴到 app/libs 中，在 jar 包上右键单击，选择 add as library，在 build.gradle(Module：App)中自动添加如下代码。

```
implementation files('libs /gson - 2.8.0.jar')
```

步骤 3：在 app/res/layout 中修改 activity_main.xml 编写布局文件，界面上放置一个文本框和按钮，然后将结果显示在另外一个文本框中，界面效果如图 10－18 所示。

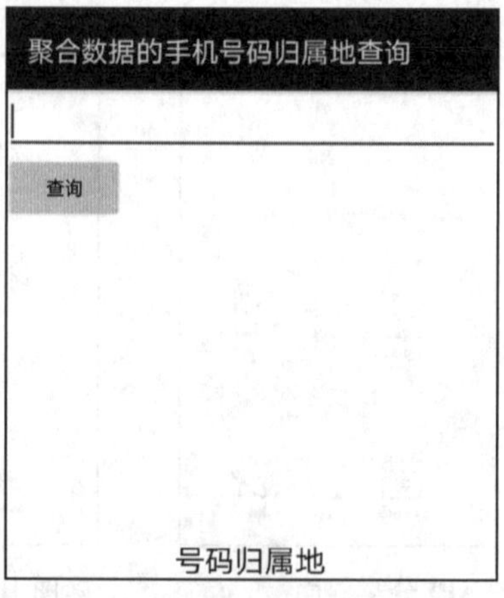

图 10－18　布局设计效果图

步骤 4：使用 GSON 需要设计对应 JSON 数据的 JavaBean 类，API 文档说明如表 10－1 所示。

表 10－1　JSON 数据格式

名称	类型	说　明
error_code	int	返回码
reason	string	返回说明
result	string	返回结果集
province	string	省份
city	string	城市(北京、上海、重庆、天津可能为空)
areacode	string	区号(部分记录可能为空)
zip	string	邮编(部分记录可能为空)
company	string	运营商

JSON 返回示例：{"resultcode":"200","reason":"Return Success!","result":{"province":"浙江","city":"杭州","areacode":"0571","zip":"310000","company":"中国移动","card":""}}。

根据返回的 JSON 数据，新建一个 Bean 将代码中表 10－1 所示的字段生成 JavaBean，包括字段和 set/get 属性，代码略。

步骤 5：修改 MainActivity，实现加载布局和控件初始化功能，代码如下。

```
public class MainActivity extends Activity implements View.OnClickListener {
    TextView textView;Button button;EditText editText;
    @Override
    protected void onCreate(Bundle savedInstanceState) {
        …………
        button.setOnClickListener(this);
        editText.addTextChangedListener(new TextWatcher() { //监听文本变化
        @Override //文本变化前
        public void beforeTextChanged(CharSequence s, int start, int count, int after){ }
       @Override
       public void onTextChanged(CharSequence s, int start, int before, int count) { String
num = s.toString();if(num.length()> 7){ } //文本变化 }
       @Override
        public void afterTextChanged(Editable s) { } //文本变化后  }); }
```

步骤 6：在 MainActivity 中，添加代码实现网络连接获得数据功能如下。

```
Handler handler = new Handler(); //使用 Handler 异步消息处理机制
 @Override
 public void onClick(View v) {
```

```
new Thread(new Runnable() {
    @Override
    public void run() { HttpURLConnection conn = null; //创建用来连接网络
        try { //拼接请求接口地址,其中 Key 后面是用户申请的 AppKey
            String url = "http: //apis. juhe. cn /mobile /get";
            String num  =  editText. getText(). toString(). trim();
            url = url + "?phone = " + num + "&key = 用户申请的 AppKey";
            ……//配置 HttpURLConnection 对象并进行网络访问获得返回数据
            final String address = parJson(response); //使用 GSON 解析 JSON 数据
            handler. post(new Runnable() { //根据返回的内容,使用子线程更新 UI
             @Override
             public void run(){textView. setText(address);} });
            } else {  throw  new  NetworkErrorException ( " response  status  is  " +
                    responseCode);}}catch (Exception e){e. printStackTrace();
            } finally {if (conn ! = null){conn. disconnect();} } } }). start();}
```

步骤 7:功能实现——把 InputStream 转换成 String。

```
private static String getStringFromInputStream(InputStream is) throws IOException {
        ByteArrayOutputStream os  =  new ByteArrayOutputStream();
        byte[ ] buffer  =  new byte[1024];   int len  =  - 1;
        while ((len  =  is. read(buffer)) ! =  - 1) {os. write(buffer, 0, len); }
        is. close();String state  =  os. toString();os. close(); return state; }}
```

步骤 8:功能实现——使用 GSON 解析 JSON 数据。

```
private String parJson(String response) {
      Gson gson = new Gson();
     Bean bean = gson. fromJson(response, Bean. class); //用 GSON 的 fromJson 解析 JSON 数据
    if("200". equals(bean. getResultcode())) { //如果结果码为 200 返回号码归属地
        Bean. ResultBean body  =  bean. getResult(); //将解析结果装入 JavaBean 中
        StringBuilder sb = new StringBuilder();
    //如果省份和城市不为空, 地址拼装省份和城市
      if(! TextUtils. isEmpty(body. getProvince())){sb. Append(body. getProvince());}
      if(! TextUtils. isEmpty(body. getCity())){sb. Append(body. getCity()); }
      return sb. toString();}else{return bean. getReason();状态码不为 200 返回错误信息}}
```

编写完成之后,运行程序,效果显示如图 10 - 19 所示。可以看到实现了通过手机号码进行归属地查询的功能。

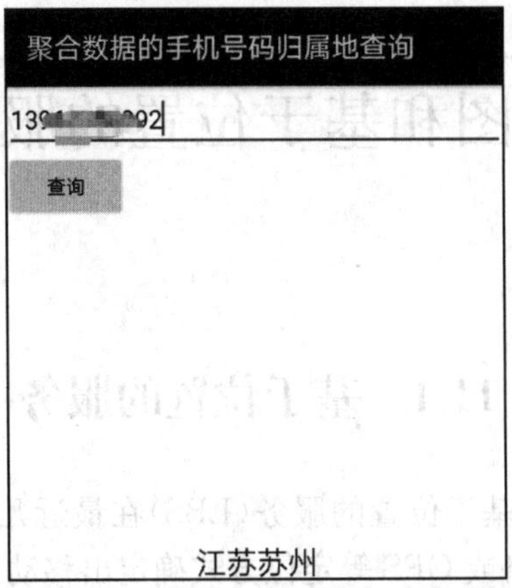

图 10－19　手机号码归属地运行结果

小结

Android 平台中提供了众多网络通信编程的接口。本章分别介绍了 WebView、HttpURLConnection、JSON 等内容，最后讲解了一个比较实用的手机号码归属地查询案例。本章内容大多需要读者进行联网学习，并具备一定的 JSP Web 开发能力。掌握本章内容有助于读者掌握客户端和服务端的通信和数据交互。

课后练习

【微信扫码】
第 10 章课后练习

第 11 章

地图和基于位置的服务

11.1　基于位置的服务

随着移动互联网的兴起，基于位置的服务(LBS)在最近几年十分火爆。它主要的工作原理就是利用无线电通信网络或 GPS 等定位方式确定出移动设备所在的位置。

基于位置的服务的核心是要先确定用户所在的位置。GPS 定位的工作原理是基于手机内置的 GPS 硬件直接和卫星交互来获取当前的经纬度信息，这种定位方式精确度非常高，但缺点是只能在室外使用，室内基本无法接收到卫星的信号。网络定位的工作原理是根据手机当前网络附近的三个基站进行测速，以此计算出手机和每个基站之间的距离，再通过三角定位确定出一个大概的位置，这种定位方式精确度一般，但优点是在室内和室外都可以使用。Android 对这两种定位方式都提供了相应的 API 支持，但是由于一些特殊原因，Google 的网络服务在中国不可访问，从而导致网络定位方式的 API 失效。而 GPS 定位虽然不需要网络，但基于以上原因，本书中不讲解 Android 原生定位 API 的用法，而是使用一些国内第三方公司的 SDK。目前国内在这一领域做得比较好的是百度和高德，本章使用百度地图的 SDK 实现 LBS 方面的一些功能。

11.2　百度地图 LBS 开发准备

11.2.1　申请百度账号

要想使用百度的 LBS 功能，需要具备几个条件：首先必须拥有一个百度账号，然后成为百度开发者，接着获得服务密钥，最后获取相关服务功能，如图 11－1 所示。

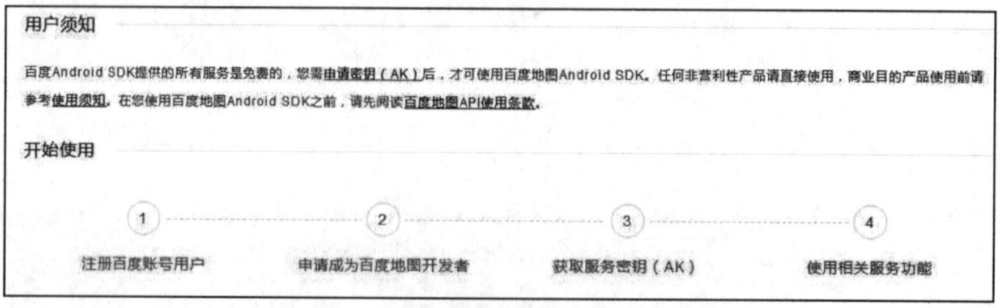

图 11－1　使用百度地图开发平台的流程

打开 https://www.baidu.com/，点击登录，然后选择立即注册，完成注册程序。注册之后，打开网址 http://lbsyun.baidu.com/，进入百度地图开放平台。

图 11－2　百度地图开放平台首页

11.2.2　创建应用获得 API Key

页面右上角为当前登录用户信息，点击菜单栏的控制台进入当前用户控制台，如图 11－3 所示。

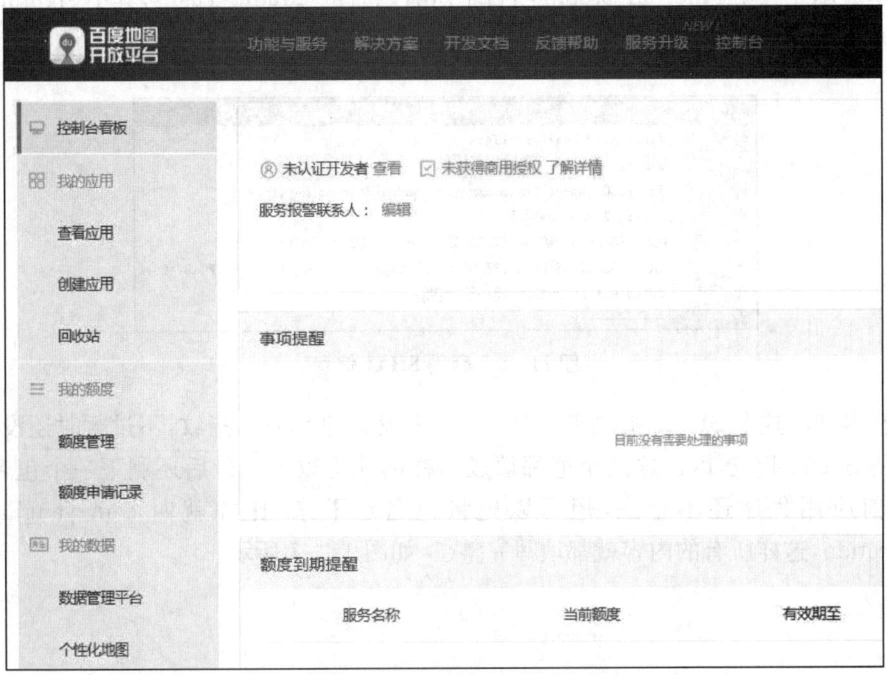

图 11－3　注册之后控制台界面

控制台是注册用户进行数据管理的平台，可以在这个控制台中创建、查看和编辑应用。点击左侧菜单栏中的创建应用，进入应用的创建界面，如图 11－4 所示。

应用名称可以随便填，应用类型选择 Android SDK，启用服务保持默认即可。那么，这个发布版 SHA1 和开发版 SHA1 又是什么呢？这是申请 API Key 所必须填写的一个字段，它是打包程序时所用签名文件的 SHA1 指纹，可以通过 Android Studio 查看到。打开 Android Studio 中的任意一个项目，点击右侧工具栏的 Gradle→项目名→：app→Tasks→Android，如图 11－5 所示。

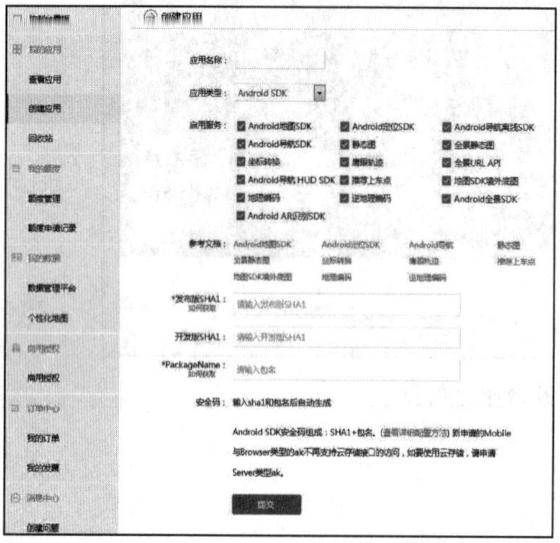

图 11－4　填写开发者信息

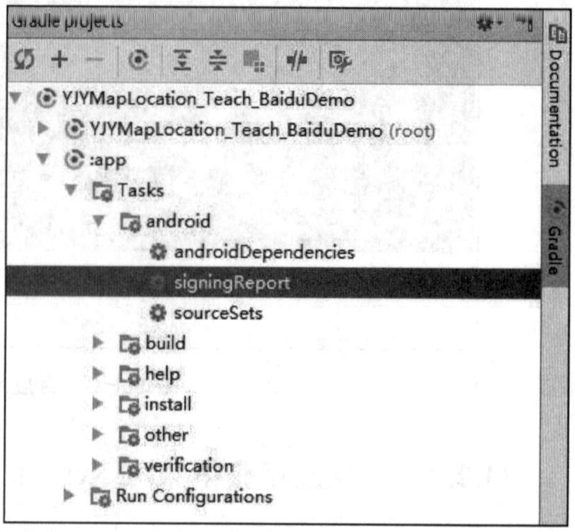

图 11－5　查看 Gradle Tasks 内容

这里展示了一个 Android Studio 项目中所有内置的 Gradle Tasks，其中 signingReport 可以用来查看签名文件信息。双击 signingReport，结果如图 11－6 所示。

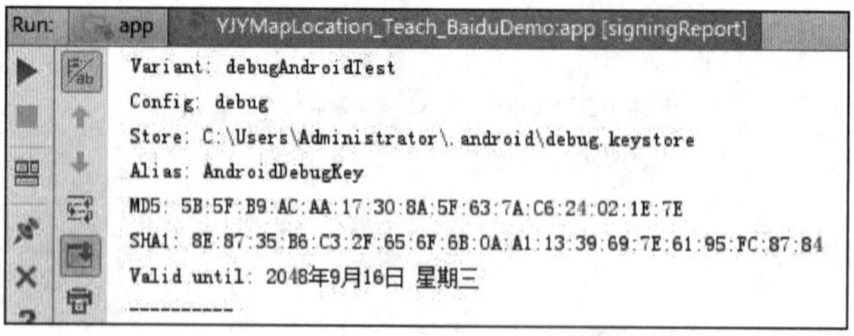

图 11－6　获得 SHA1 信息

现在得到的这个 SHA1 指纹实际上是一个开发版的 SHA1 指纹，不过暂时还没有一个发布版的 SHA1 指纹，因此这两个值都填成一样的就可以了。最后还剩下一个包名选项，虽然目前应用程序还不存在，但可以先将包名定下来，比如就叫 com.yangjianyong.baidulocation，这样所有的内容就都填写完整了，如图 11－7 所示。

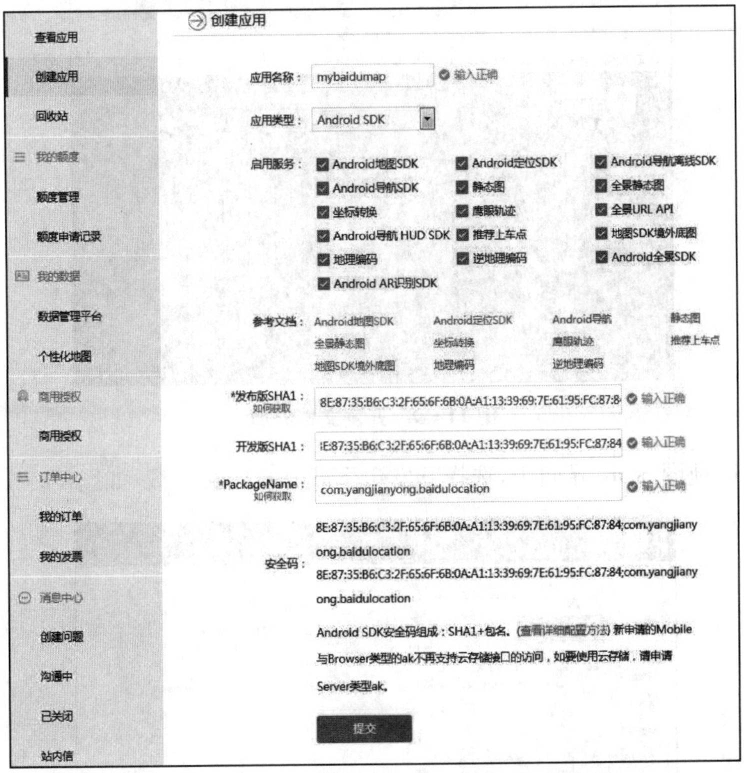

图 11－7 填写完整的应用信息

接下来点击提交，如果看到已创建的应用如图 11－8 所示就表示创建成功了。其中 BIxLmLN06XK0CakcZPSkOGet0GaKoqOo 就是申请到的 API Key，有了它就可以进行后续的 LBS 开发工作了。

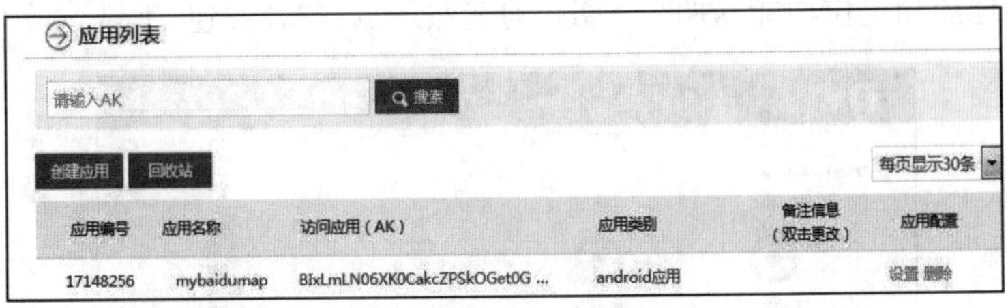

图 11－8 查看已创建的应用

11.2.3 LBS SDK 的获取和使用

在开始编码之前，需要先将百度 LBS 开放平台的 SDK 准备好，打开网址 http://lbsyun.baidu.com。点击开发文档，选择 Android 开发中的 Android 地图 SDK，如图 11－9 所示。

图 11 - 9　开发文档界面

在 Android 地图 SDK 界面中，选择开发包下载→自定义下载，如图 11 - 10 所示。

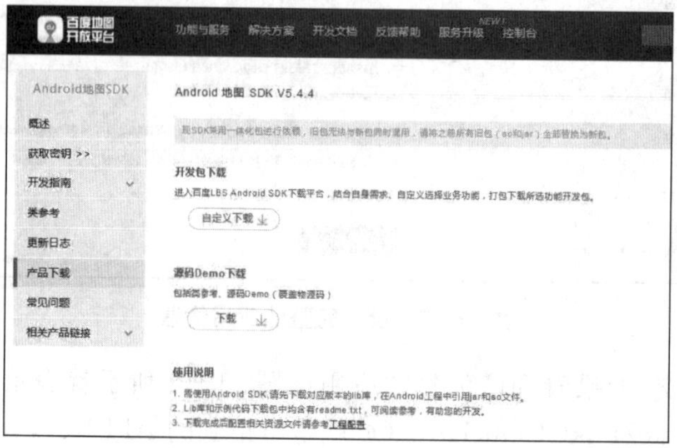

图 11 - 10　Android 地图 SDK 下载

找到基础地图和定位功能这两个 SDK，勾选然后点击“开发包”下载，如图 11 - 11 所示。

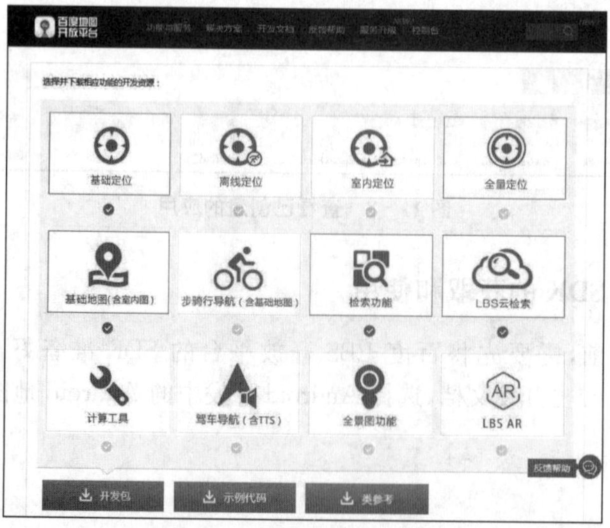

图 11 - 11　下载 SDK 界面

下载完成后解压压缩包，其中会有一个 libs 目录，里面的内容就是开发需要的 API。

libs 目录下的内容分为两部分：BaiduLBS Android.jar 文件是 Java 层要使用到的；其他子目录下的 so 文件是 Native 层要用到的。so 文件是用 C/C ++语言进行编写，然后再用 NDK 编译出来的。当然这里并不需要编写 C/C ++的代码，因为百度都已经做好了封装，但是需要将 libs 目录下的每一个文件都放置到正确的位置。做好了上面的准备工作，可以开始进行基于百度地图的位置的服务开发了。

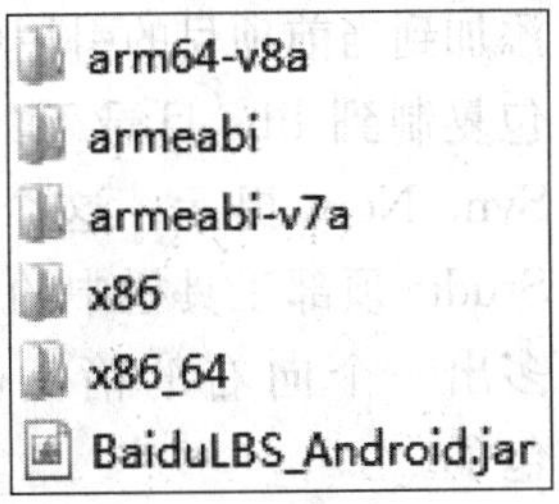

图 11－12　压缩包 libs 目录下的内容

11.3　基于百度地图 LBS 的开发

【微信扫码】案例 11－1 相关文件

案例 11－1：基于百度地图位置的服务开发

新建一个项目，命名为 MapLocation_Teach_BaiduDemo。

11.3.1　项目开发包和全局文件配置

首先要将 LBS SDK 中的 jar 包导入项目中。在 Android Studio 的 IDE 中，点击左上角的视图，切换到 Project 模式，如图 11－13 所示。首先观察一下当前的项目结构，你会发现 App 模块下面有一个 libs 目录，这里就是用来存放所有的 jar 包的，将 BaiduLBS_Android.jar 复制到这里，如图 11－14 所示。

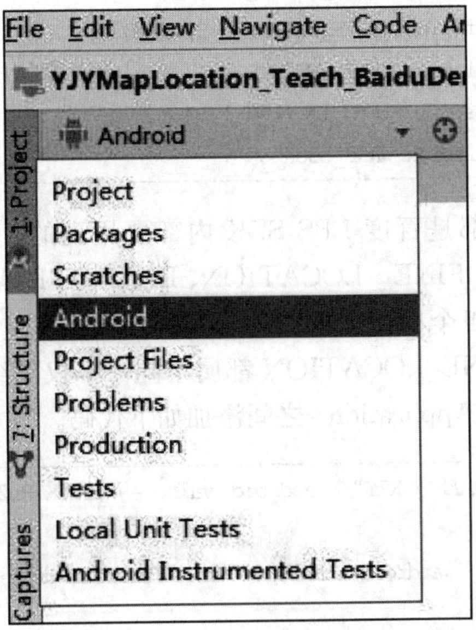

图 11－13　开发视图模式切换

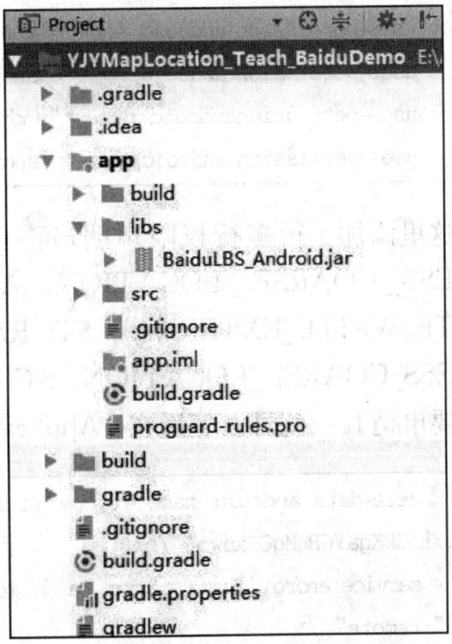

图 11－14　导入 jar 包到项目 libs 目录中

每个项目的 app/build.gradle 文件中，在 dependencies{}部分都会声明：implementation fileTree(include: ['*.jar'], dir: 'libs')。这表示将 libs 目录下所有以 jar 后缀结尾的文件

添加到当前项目的引用中。但由于刚才的操作是直接将 jar 包复制到 libs 目录下的，没有修改 gradle 文件，不会弹出 Sync Now 提示。这个时候必须手动点击一下 Android Studio 顶部工具栏中的 Sync 按钮，当 libs 目录下的 jar 文件多出一个向右的箭头，表示项目已经能引用到这些 jar 包了。

为了确保 jar 包能被项目引用，还需要在 app/build.gradle 的 dependencies{}中配置代码：implementation files('libs/BaiduLBS_Android.jar')。然后点击 Android Studio 顶部工具栏中的 Sync 按钮，刷新项目引用。接着，展开 src/main 目录，右击该目录 New/Directory，创建名为 jniLibs 的目录，把压缩包里的其他所有目录直接复制到这里，如图 11-15 所示。这样就完成了开发包的配置。

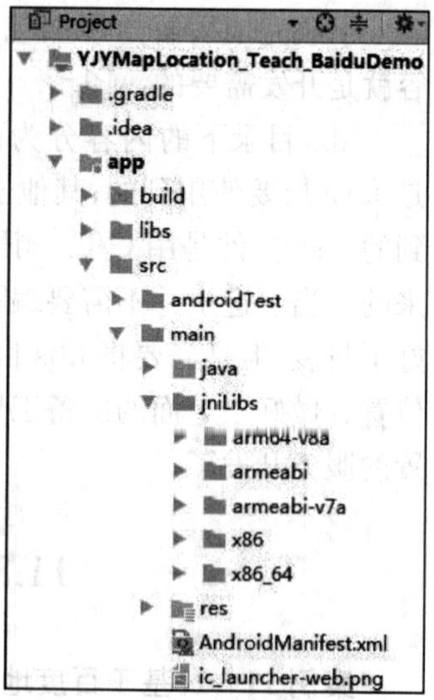

图 11-15　将 so 文件放置到 jniLibs 目录中

接着修改 AndroidManifest.xml 文件的代码，添加权限声明。

```
<uses-permission android:name="android.permission.ACCESS_COARSE_LOCATION" />
<uses-permission android:name="android.permission.ACCESS_FINE_LOCATION" />
<uses-permission android:name="android.permission.ACCESS_WIFI_STATE" />
<uses-permission android:name="android.permission.ACCESS_NETWORK_STATE" />
<uses-permission android:name="android.permission.CHANGE_WIFI_STATE" />
<uses-permission android:name="android.permission.READ_PHONE_STATE" />
<uses-permission android:name="android.permission.WRITE_EXTERNAL_STORAGE" />
<uses-permission android:name="android.permission.INTERNET" />
<uses-permissionandroid:name="android.permission.MOUNT_UNMOUNT_FILESYSTEMS" />
<uses-permission android:name="android.permission.WAKE_LOCK" />
```

这里添加了很多行权限声明，每一个权限都是百度 LBS SDK 内部要用到的。其中 ACCESS_COARSE_LOCATION、ACCESS_FINE_LOCATION、READ_PHONE_STATE、WRITE_EXTERNAL_STORAGE 这四个权限是需要运行时处理的，不过由于 ACCESS_COARSE_LOCATION、ACCESS_FINE_LOCATION 都属于同一个权限组，两者只要申请其一就可。然后在<Application></Application>之间添加如下代码。

```
<meta-data android:name="com.baidu.lbsapi.API_KEY"   android:value="irMYKYRn2sNv
XhlCBKgaY6BnMqGRanwo" />
<service android:name="com.baidu.location.f"   android:enabled="true" android:process=
":remote" />
```

meta-data 标签 android:name 部分必须填 com.baidu.lbsapi.API_KEY。android:value 部分填入在前面申请到的 API Key。<service></service>为注册一个 LBS SDK 中的服务。

11.3.2 界面设计和地图显示

修改 activity_main.xml 代码，在布局文件中新放置了一个 MapView 控件，并让它填充满整个屏幕。MapView 是由百度提供的自定义控件，使用它的时候需要将完整的包名加上。

在界面的左上角放置一个用于定位的按钮，通过控制 android:layout_marginTop 和 android:layout_marginLeft，让其浮在 MapView 上位置，代码如下。

```
< com.baidu.mapapi.map.MapView  android:clickable = "true" ……  />
< Button android:id = "@ + id /location_btn"  ……  />
```

根据布局文件，修改 MainActivity，主要在 onCreate()方法中添加如下代码。

```
public class MainActivity extends AppCompatActivity {
    private MapView mapView;
    @Override
    protected void onCreate(Bundle savedInstanceState) {
        super.onCreate(savedInstanceState);
        SDKInitializer.initialize(getApplicationContext());
        setContentView(R.layout.activity_main);
        mapView = findViewById(R.id.bmapView); //获取到了 MapView 的实例  }
```

SDKInitializer.initialize(getApplicationContext())为调用 SDKInitializer 的 initialize()方法进行初始化操作，initialize()方法接收一个 Context 参数，调用 getApplicationContext()方法获取一个全局的 Context 参数并传入。这个操作必须在 setContentView()方法之前，否则会出错。接着，在 MainActivity 中重写 onResume()、onPause()和 onDestroy()这三个方法，在这里对 MapView 进行管理，实现地图生命周期管理。

```
@Override
protected void onResume() {super.onResume();mapView.onResume();}
@Override
protected void onPause() {super.onPause();mapView.onPause();}
@Override
protected void onDestroy() {super.onDestroy();mapView.onDestroy();}
```

现在运行一下程序，百度地图就应该成功显示出来了，如图 11－16 所示。

11.3.3 系统权限

上面通过简单的几行代码，成功地将地图显示出来了，但这是一张默认的地图，显示的是北京市中心的位置，而要想获得更加精细的地图信息，比如自己所在位置的周边环境，就需要使用系统的一些权限，比如 GPS、系统读写和读取电话状态等。在 AndroidManifest.xml 已经声明了权限，在程序运行中需要动态申请。在 MainActivity 中添加 getPermissions()方法，代码如下。

```
private void getPermissions() {
    List < String > permissionList = new ArrayList <>(); //创建空的 List 集合,存放权限
    if (ContextCompat.checkSelfPermission(MainActivity.this, Manifest.permission.ACCESS_
FINE_LOCATION) ! = PackageManager.PERMISSION_GRANTED) {
     permissionList.add(Manifest.permission.ACCESS_FINE_LOCATION);}
    …… //依次判断权限有没有被授权,如果没被授权就添加到 List 集合中
    if (!permissionList.isEmpty()) { //将 List 转换成数组
      String [ ] permissions = permissionList.toArray(new String[permissionList.size()]);
      ActivityCompat.requestPermissions(MainActivity.this, permissions, 1); //一次性申请
    } else { requestLocation(); } }
```

在 MainActivity 中添加 onRequestPermissionsResult()方法实现对授权结调。代码如下。

```
@Override
public void onRequestPermissionsResult ( int requestCode, String [ ] permissions, int [ ]
grantResults) {
  switch (requestCode) {
    case 1: if (grantResults.length > 0) { for (int result : grantResults) {
       if (result ! = PackageManager.PERMISSION_GRANTED){ Toast.makeText(this, "必须同意
所有权限才能使用本程序", Toast.LENGTH_SHORT).show();
        finish(); return;}}  requestLocation();
    } else {Toast.makeText(this, "发生未知错误", Toast.LENGTH_SHORT).show();
        finish(); } break;  default:   }  }
```

onRequestPermissionsResult()方法中对权限申请结果的逻辑处理和之前有所不同,通过一个循环将申请的每个权限都进行了判断,如果有任何一个权限被拒绝,就直接调用 finish()方法关闭当前程序,只有当所有权限都被用户同意了,才会调用 requestLocation()方法开始地理位置定位。修改 MainActivity,在 onCreate()方法中,添加权限申请的代码如下。

```
public class MainActivity extends AppCompatActivity {
    @Override
    protected void onCreate(Bundle savedInstanceState) {……   getPermissions(); }
```

当运行程序时,会出现权限的申请界面,如图 11 - 17 所示。

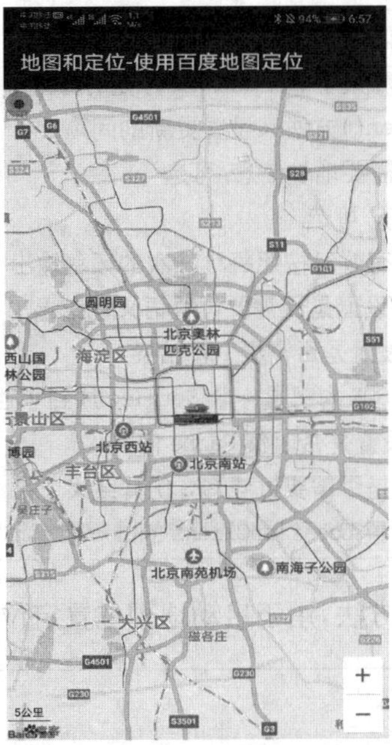

图 11－16　百度地图的显示

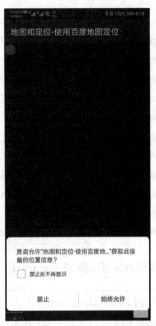

图 11－17　动态权限申请

11.3.4　地图的定位模式

前面介绍了 Android 的有两种定位方式，第一种是通过 GPS 定位，第二种是通过网络定位。本案例我们使用 GPS 高精确度模式定位。修改 MainActivity，在 onCreate()方法中添加如下代码。

```
public class MainActivity extends AppCompatActivity {
    public LocationClient mLocationClient;
    @Override
    protected void onCreate(Bundle savedInstanceState) {
      ……getPersimmions();
     //创建一个 LocationClient 的实例，构建函数接收 Context 参数
     mLocationClient = new LocationClient(getApplicationContext());
     //注册定位监听器，当获取到位置信息的时候，就会回调这个定位监听器.
     mLocationClient.registerLocationListener(new MyLocationListener());}
```

在 MainActivity 中添加 initLocation()，设置 LBS SDK 定位模式进行设置，代码如下。

```
private void initLocation(){
  LocationClientOption option = new LocationClientOption();
   //设置定位模式为高精确度
  option.setLocationMode(LocationClientOption.LocationMode.Hight_Accuracy);
```

```
option.setCoorType("bd09ll"); //设置返回的定位结果坐标系,可选
int span = 1000; //设置发起定位请求的间隔,大于等于 1000ms 才是有效的
option.setScanSpan(span); //调用 setScanSpan()方法来设置更新的间隔
option.setIsNeedAddress(true); //设置是否需要地址信息,可选
option.setOpenGps(true); //设置是否使用 GPS,可选
option.setLocationNotify(true); //设置是否有效时按照每秒一次的频率输出 GPS 结果
option.setIsNeedLocationDescribe(true); //设置是否需要位置语义化结果
option.setIsNeedLocationPoiList(true); //设置是否需要 Poi 结果
option.setIgnoreKillProcess(false); //定位 SDK 内部是一个 Service 放到了独立进程
option.SetIgnoreCacheException(false); //设置是否收集 Cache 信息
option.setEnableSimulateGps(false); //设置是否需要过滤 GPS 仿真结果
mLocationClient.setLocOption(option) //设置 LocationClient 的 LocOption; }
```

在 MainActivity 中,新增 requestLocation()方法,用于启动地理位置定位功能。

```
private void requestLocation() {
   initLocation(); //配置 LocationClient 的参数
   mLocationClient.start();}
```

使用 LocationClient 的 start()方法就能开始定位了。定位的结果会回调到监听器 mLocationClient.registerLocationListener(new MyLocationListener())中。

两次修改前面编写的 onDestroy()方法,增加代码 mLocationClient.stop()。在活动被销毁时,调用 LocationClient 的 stop()方法来停止定位,不然程序会在后台不停地进行定位,从而严重消耗手机的电量。

11.3.5 地图的位置移动

利用得到的经纬度信息,在地图中快速移动到自己的位置。

百度 LBS SDK 的 API 中提供了一个 BaiduMap 类,它是地图的总控制器,可以对地图进行各种操作,比如设置地图的缩放级别以及将地图移动到某一个经纬度上。百度地图将缩放级别的取值范围限定在 3 到 19,其中小数点位的值也是可以取的,值越大,地图显示的信息就越精细。移动地图到某一个经纬度上就需要借助 LatLng 类了,主要就是用于存放经纬度值,它的构造方法接收两个参数,第一个参数是纬度值,第二个参数是经度值。之后调用 MapStatusUpdateFactory 的 newLatLng()方法将 LatLng 对象传入,newLatLng()方法返回的也是一个 MapStatusUpdate 对象,再把这个对象传入 BaiduMap 的 animateMapStatus()方法中,就可以将地图移动到指定的经纬度上了。

修改 MainActivity,在 onCreate()方法中添加如下代码。

```
public class MainActivity extends AppCompatActivity {
    …………
    private BaiduMap baiduMap;private boolean isFirstLocate = true;
    @Override
```

```
    protected void onCreate(Bundle savedInstanceState) {
        …………
      baiduMap = mapView.getMap();}
```

调用 MapView 的 getMap()方法就能获取 BaiduMap 的实例。

在 MainActivity 中添加 navigateTo()方法，作用是移动地图到设备所在位置，代码如下。

```
private void navigateTo(BDLocation location) {
   if (isFirstLocate) { //使用 isFirstLocate 变量,变量的作用是为了防止多次调用内置的方
法,只需要在程序第一次定位的时候调用一次就可以
     Toast.makeText(this, "位置标注" + location.getAddrStr(),Toast.LENGTH_SHORT).show();
      //将 BDLocation 对象中的地理位置信息取出并封装到 LatLng 对象中
     LatLng ll = new LatLng(location.getLatitude(),location.getLongitude());
      //调用 MapStatusUpdateFactory 的 newLatLng()方法并将 LatLng 对象传入
    MapStatusUpdate update = MapStatusUpdateFactory.newLatLng(ll);
    baiduMap.animateMapStatus(update);
  //将返回的 MapStatusUpdate 对象作为参数传入 BaiduMap 的 animateMapStatus()方法当中。为
了让地图信息可以显示得更加丰富一些,将缩放级别设置成 16
   update = MapStatusUpdateFactory.zoomTo(16f);
     baiduMap.animateMapStatus(update);
   isFirstLocate = false; } }
```

在 MyLocationListener 的 onReceiveLocation()方法中调用 navigateTo()，移动地图到设备所在位置。

```
public class MyLocationListener implements BDLocationListener {
    @Override
    public void onReceiveLocation(BDLocation location) {
     …………
    if (location.getLocType() = = BDLocation.TypeGpsLocation || location.getLocType() ==
BDLocation.TypeNetWorkLocation) {navigateTo(location); }}}
```

11.3.6　位置信息的转换

虽然可以通过设备所在的经纬度移动到设备所在的位置，但这种经纬度的值一般人是根本看不懂的，相信谁也无法立刻答出南纬 30 度、东经 120 度是什么地方。我们除了想看当前的经纬度，还想看看其他更丰富的位置信息，比如在什么路或者地方。

百度 LBS SDK 在这方面提供了非常好的支持，只需要进行一些简单的接口调用就能得到当前位置下各种丰富的信息，在监听器 MyLocationListener 的 onReceiveLocation 方法中，利用传入的 BDLocation 参数，调用 getXXX()得到相应的信息，代码如下。

```
public class MyLocationListener implements BDLocationListener {
    @Override
    public void onReceiveLocation(BDLocation location) {
    ……
    StringBuilder currentPosition = new StringBuilder();
    currentPosition.Append("纬度:").Append(location.getLatitude()).Append("\n");
    ……//根据上面代码完成经线 getLongitude()、国家 getCountry()、省 getProvince()、市
getCity()、区 getDistrict()、街道.getStreet()等信息获取
    currentPosition.Append("定位方式:");
    if (location.getLocType() == BDLocation.TypeGpsLocation) {
            currentPosition.Append("GPS");
      } else if (location.getLocType() == BDLocation.TypeNetWorkLocation) {
          currentPosition.Append("网络");  }
    Toast.makeText(MainActivity.this,currentPosition,Toast.LENGTH_LONG).show();    }}
```

11.3.7 显示当前位置的光标

通常情况下,手机地图上都会有一个小光标,用于显示设备当前所在的位置,并且如果设备正在移动的话,这个光标也会跟着一起移动。

百度 LBS SDK 中提供了一个 MyLocationData.Builder 类,这个类是用来封装设备当前所在位置的,只需将经纬度信息传入这个类的相应方法当中就可以了。

MyLocationData.Builder 类还提供了一个 build()方法,把要封装的信息都设置完成后,只需要调用 build()方法,就会生成一个 MyLocationData 实例,再将这个实例传入 BaiduMap 的 setMyLocationData()方法中,就可以让设备当前的位置显示在地图上。继续对现有代码进行扩展,让这个光标能够显示在地图上。

在 MainActivity 的 onCreate()方法中添加如下代码。

```
public class MainActivity extends AppCompatActivity {
    @Override
    protected void onCreate(Bundle savedInstanceState) {
        ……baiduMap.setMyLocationEnabled(true);}
```

在 navigateTo()方法中,新增如下代码。

```
private void navigateTo(BDLocation location) {
    if (isFirstLocate) {  ……  }
   //添加 MyLocationData 的构建逻辑
    MyLocationData.Builder locationBuilder = new MyLocationData.Builder();
    locationBuilder.latitude(location.getLatitude()); //封装 Location 中包含的经度
    locationBuilder.longitude(location.getLongitude()); //封装 Location 中包含的经度
     //MyLocationData 设置到 BaiduMap 的 setMyLocationData()方法中
    MyLocationData locationData = locationBuilder.build();
    baiduMap.setMyLocationData(locationData);}
```

这段代码必须写在 isFirstLocate 语句外面，因为地图移动位置只需要在第一次定位的时候执行，但是在地图上显示的光标位置是随着设备的移动而实时改变的。

在 onDestroy()方法中添加如下代码：baiduMap.setMyLocationEnabled(false)；在程序退出的时候，要将此功能给关闭掉。

11.3.8　定位按钮功能实现

在手机地图中，通常会有一个按钮用来重新定位当前位置，因为功能代码在前面都已实现，按钮点击事件中只要调用 requestLocation()即可，代码如下。

```
public class MainActivity extends AppCompatActivity {
  ………… public Button btn_location;
  @Override
  protected void onCreate(Bundle savedInstanceState) {
    ………… btn_location = findViewById(R.id.location_btn);
   btn_location.setOnClickListener(new View.OnClickListener() {
   @Override
   public void onClick(View v){ isFirstLocate = true; requestLocation(); //二次定位 }});}
```

编写完成后，运行程序，效果如图 11－18 所示。用户可以清晰地看出自己当前的位置。

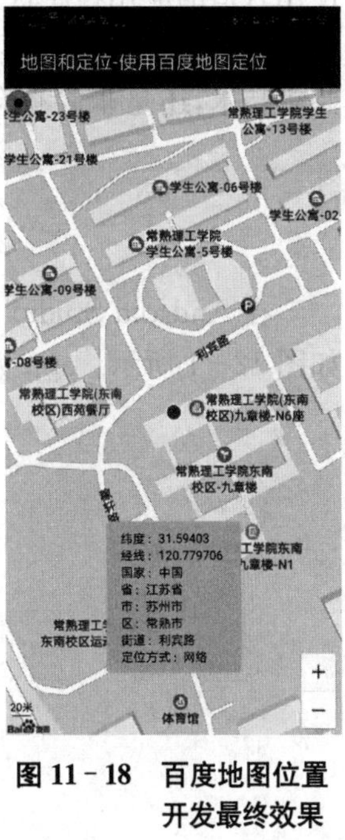

图 11－18　百度地图位置开发最终效果

小结

关于百度 LBS SDK 的用法就介绍到这里。如果本章案例程序能运行出来，已经算是成功入门了。如果想要更加深入地研究百度 LBS 的各种用法，可以到官方网站上参考开发指南，使用官网的开发指南来进行学习也是非常重要的，因为官方文档永远是最新的。

课后练习

【微信扫码】
第 11 章课后练习

第 12 章

Android 项目开发:文件管理 App

通过前面的学习,我们掌握了 Android 开发的很多知识,学习这些知识的最终目的是为了能够开发一个功能相对完整的 App 程序。本章通过真实的项目来检验学习的效果,我们模仿安卓应用市场上的文件管理 App,开发一个基于 Android 平台的文件管理系统程序。

12.1 项目需求分析

要做一个 App,首先要对程序进行需求分析,想一想这样的应用程序应该具备什么样的功能,然后将这些功能全部整理出来。经过项目需求分析,按照日常生活所需文件要求,基于 Android 平台下的文件管理系统,具体的功能需求:① 能够打开任一文件夹,并浏览目录下的文件及文件夹。② 能够在任何一个位置,返回上一级目录或者返回根目录。③ 能够使用合适的应用程序打开文件。④ 能够对文件目录或者文件进行复制及粘贴操作。虽然看上去只有几个功能,但是需要用到 Activity、View、Adapter、事件监听、数据存储、权限控制等技术,还是非常考验综合应用能力的。

12.2 项目的程序结构

在 Android Studio 中新建一个项目,命名为 Project_Teach_FileApp。项目的程序结构如图 12-1 所示。app 为项目根目录,它下面的几个目录作用如下。

图 12-1 项目的程序架构

manifests:主要文件为 AndroidManifest.xml,它是整个 Android 项目配置文件,除了在文件中给应用程序添加权限声明之外,Android 系统的四大组件都需要在这个文件里注册。

java：存放 Java 代码。

res：存放项目的资源文件，根据不同资源类型分为：drawable 存放项目中所用的图片；layout 存放项目的布局文件；mipmap 存放项目的应用图标；values 存放项目中使用的字符串、样式、颜色等文件；xml 存放项目中使用的 XML 文件。

12.3　界面设计和资源文件

在 Android 程序中，界面占了工作的很大一部分，界面的好坏决定了项目的品质外观。在本项目中，界面分为主界面和详细界面两个部分。

12.3.1　主界面设计

主界面的布局文件命名为 activity-main，布局的核心结构代码如下。

```
< ListView   android: id = "@ + id /androidlist"……  />
< LinearLayout android: id = "@ + id /linearLayout"   ……>
    < LinearLayout   android: id = "@ + id /laout_paste"   >
        < ImageView android: src = "@drawable /menu_paste" />
        < TextView android: text = "@string /menu_text_paste" />
    < /LinearLayout >
    < LinearLayout android: id = "@ + id /laout_exit"   >
        < ImageView android: src = "@drawable /menu_exit" />
        < TextView   android: text = "@string /menu_text_exit"  />
    < /LinearLayout >
< /LinearLayout >
```

首先，在界面上放一个 ListView 布局文件用于显示当前目录的内容，宽度和高度设置成 match-parent 占满整个空间。设置条目分割线图片（存放在 drawable 中的名为 line 的图片），然后在容器中居左顶部对齐，紧贴父元素结束的位置开始。其次下方放一个线性布局文件用于显示菜单。大线性布局中包含两个垂直的线性布局文件，构成底部的菜单，分别为粘贴和退出按钮。通过将 layout_width 设置为 0 和 layout_weight 设置为 1 的比例分配方式，确保菜单均匀分布。ListView 控件置于 LinearLayout 控件之上，效果如图 12 - 2 所示。

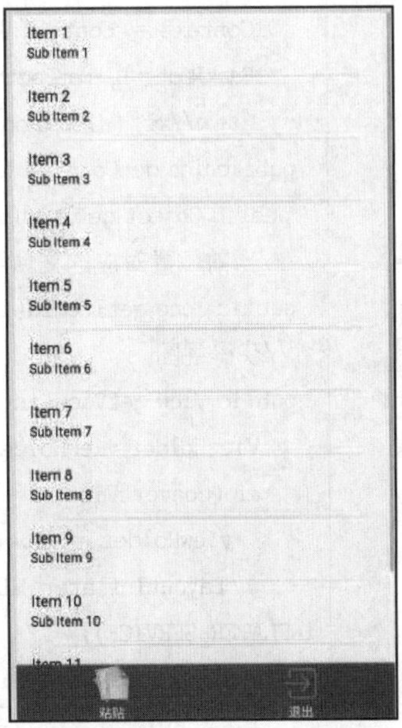

图 12 - 2　主界面设计效果

12.3.2　详细信息界面设计

小布局文件用于显示实际文件信息，在 LinearLayout 布局放置 ImageView 显示图标和 TextView 显示文件名，靠左居中对齐，效果如图 12 - 3 所示，代码如下。

```
<LinearLayout   ...... >
    <ImageView android:id = "@ + id /image_list_childs" />
    <TextView android:id = "@ + id /text_list_childs" android:ellipsize = "marquee"
        android:marqueeRepeatLimit = "marquee_forever"   ......  />
</LinearLayout>
```

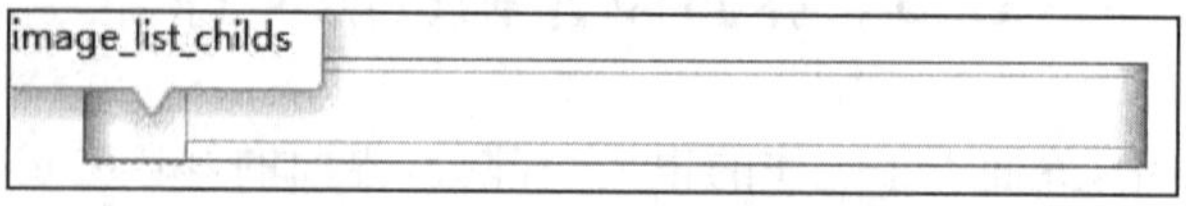

图 12-3　文件详细信息显示布局

12.4　文件适配器核心功能实现

程序启动之后，首先要显示文件系统的目录结构。对于这种需要按照条目进行数据获取和显示的需求，Android 系统有非常好的解决方式，就是在第 5 章中介绍的适配器技术。这里我们使用 BaseAdapter 来完成此功能。

在 app/java/程序包中，新建一个 Java 程序，命名为 FileAdapter，这个类继承于基础适配器 BaseAdapter 并进行扩展，代码如下。

```
public class FileAdapter extends BaseAdapter {
  private Bitmap lBackRoot;
  ...... //参照上定义其他各种文件的图标
  private List<String> lFileNameList;  //文件名列表
  private List<String> lFilePathList;  //文件对应的路径列表
  public FileAdapter(Context context, List<String> fileName, List<String> filePath) {
    lContext = context;lFileNameList = fileName;lFilePathList = filePath;
    lBackRoot = BitmapFactory.decodeResource(lContext.getResources(),R.drawable.back_to_root);
    ...... //参照 lBackRoot 获得其他各种文件类型的图标 }
  public int getCount() {return lFilePathList.size();} //获得文件的总数
  public Object getItem(int position){return lFileNameList.get(position);}
  //当前位置对应的文件名
  public long getItemId(int position) {return position;} //获得当前的位置
  //获得视图
  public View getView(int position, View convertView, ViewGroup viewgroup){
     ViewHolder viewHolder = null;
     if (convertView == null) {
       viewHolder = new ViewHolder();
       LayoutInflater mLI = (LayoutInflater) lContext.getSystemService(Context.LAYOUT_
INFLATER_SERVICE);
      convertView = mLI.inflate(R.layout.list_child, null);
```

```
        //获取列表布局界面元素
        viewHolder.lIV = (ImageView)convertView.findViewById(R.id.image_list_childs);
        viewHolder.lTV = (TextView)convertView.findViewById(R.id.text_list_childs);
        convertView.setTag(viewHolder); //将每一行的元素集合设置成标签
    }else{viewHolder = (ViewHolder) convertView.getTag(); //获取视图标签}
    File mFile = new File(lFilePathList.get(position).toString());
    if (lFileNameList.get(position).toString().equals("BacktoRoot")) {
        viewHolder.lIV.setImageBitmap(lBackRoot);viewHolder.lTV.setText("返回根目录");
        }else if (lFileNameList.get(position).toString().equals("BacktoUp")){
        viewHolder.lIV.setImageBitmap(lBackUp);viewHolder.lTV.setText("返回上一级");
       }else{String fileName = mFile.getName();viewHolder.lTV.setText(fileName);
        if (mFile.isDirectory()){viewHolder.lIV.setImageBitmap(lFolder);
        }else{String fileEnds = fileName.substring(fileName.lastIndexOf(".") + 1,fileName.
length()).toLowerCase(); //取出文件后缀名并转成小写
        viewHolder.lIV.setImageBitmap(getApplicationBitmap(fileEnds));}}
          return convertView;}
  public Bitmap getApplicationBitmap(String fileEnds){
    Bitmap bitmap;
    switch (fileEnds){
      case "m4a":case "mp3":case "mid":case "ogg":case "wav"":bitmap = lVideo;break;
      …… //参照上面对其他图标变量进行赋值
      default: bitmap = lOthers;} return bitmap;}
    class ViewHolder{ImageView lIV;TextView lTV;} //存储列表各行元素的图片和文本}
```

12.5　主程序 MainActivity 结构

在 MainActivity 中，实现权限控制、控件初始化、文件目录的显示和浏览、应用程序打开能、复制和粘贴等功能。

12.5.1　控件初始化和功能实现

主界面上有显示文件列表的线性布局控件和用于复制粘贴的按钮控件，控件在声明之后需要进行初始化和功能实现，在 MainActivity 中，添加如下模块。

```
private void initView() {
  exit_layout = findViewById(R.id.laout_exit); exit_layout.setOnClickListener(this);
  paste_layout = findViewById(R.id.laout_paste);paste_layout.setOnClickListener(this);}
```

initView()模块用于获取相关控件。代码中通过 findViewById 初始化控件之后，设置对应的单击相应事件，通过 onClick(View v)来处理，用于底部菜单单击的功能实现。

```
public void onClick(View v) {
    switch (v.getId()) {
```

```
        case R.id.laout_paste:onPaste();break;
        case R.id.laout_exit:MainActivity.this.finish();break;}}
```

MainActivity 直接继承了 View.OnClickListener，可以直接使用 onClick(View v)来处理单击事件。在事件中通过控件的 ID 值来区分粘贴和退出控件，并执行不同的操作。

12.5.2 文件浏览和打开功能实现

在文件浏览的界面上，有返回根目录和返回上一级两个按钮控件。在 MainActivity 中，新建一个模块，initAddBackUp 用于按钮控件的初始化。

```
private boolean isAddBackUp = false;
private void initAddBackUp(String filePath, String phone_sdcard) {
    if (!filePath.equals(phone_sdcard)) {
        lFileName.add("BacktoRoot"); //列表项的第一项设置为返回根目录
        lFilePaths.add(phone_sdcard);
        lFileName.add("BacktoUp"); //列表项的第二项设置为返回上一级
        lFilePaths.add(new File(filePath).getParent()); //设置返回上一级按钮的路径为回
到当前目录的父目录
        isAddBackUp = true; //将添加返回按键标识位置为 true}}
```

在 MainActivity 中，新建一个模块，命名为 initFileListInfo(String filePath)用于文件夹内的文件和文件夹的获取，实现效果如图 12－4 所示。

图 12－4　文件信息显示

图 12－5　使用应用程序打开程序

```
public static String lCurrentFilePath = ""; //当前路径信息
private void initFileListInfo(String filePath) {
    isAddBackUp = false;lCurrentFilePath = filePath; File mFile = new File(filePath);
```

```
lFileName = new ArrayList<String>();lFilePaths = new ArrayList<String>();
File[] mFiles = mFile.listFiles(); //获得指定路径下的所有文件/文件夹
initAddBackUp(filePath, lSDCard);
//将所有文件信息添加到集合中,文件信息包括文件名和文件路径
for (File mCurrentFile : mFiles) {
    lFileName.add(mCurrentFile.getName());lFilePaths.add(mCurrentFile.getPath());}
//使用文件适配器,显示集合中的全部文件和文件夹
setListAdapter(new FileAdapter(MainActivity.this, lFileName, lFilePaths));}
```

在 MainActivity 中，新建一个模块，命名为 openFile，打开文件和文件夹操作。

```
private void openFile(File file) {
    //根据文件信息匹配,获取当前文件名称,再根据文件名称匹配 MIME
    MimeTypeMap mime = MimeTypeMap.getSingleton();
    String ext = file.getName().substring(file.getName().lastIndexOf(".") + 1);
    String type = mime.getMimeTypeFromExtension(ext);
    try {
      Intent intent = new Intent(); //设置 intent 并启动跳转
      intent.setAction(Intent.ACTION_VIEW);
      if (Build.VERSION.SDK_INT >= Build.VERSION_CODES.N) {
        intent.setFlags(Intent.FLAG_GRANT_READ_URI_PERMISSION);
        Uri contentUri = FileProvider.getUriForFile(this,"com.yangjianyong.myfilecopyandpaste",
file);
        intent.setDataAndType(contentUri, type);
        } else { intent.setDataAndType(Uri.fromFile(file), type);}
        startActivityForResult(intent, ACTIVITY_VIEW_ATTACHMENT);
    } catch (ActivityNotFoundException anfe) {
   Toast.makeText(this, "没有可以打开当前文件的应用!",Toast.LENGTH_LONG).show();}}
```

在 MainActivity 中，新建一个模块 onListItemClick，当用户单击某个文件时候，根据类型筛选，可以选择继续打开文件夹或者使用应用程序打开。

```
@Override
protected void onListItemClick(ListView listView, View view, int position, long id) {
   final File mFile = new File(lFilePaths.get(position));
   if (mFile.canRead()){ //如果文件且可读,就打开文件
     if (mFile.isDirectory()) { //如果是文件夹,则直接进入该文件夹,查看文件目录
         initFileListInfo(lFilePaths.get(position));
     }else{ openFile(mFile); //如果是文件根据筛选结果使用对应程序打开}
   }else{ Toast.makeText(MainActivity.this, "对不起,您的访问权限不足!", Toast.LENGTH_
SHORT).show();}}
```

12.5.3 复制功能实现

复制功能实现的步骤:① 首先判断该文件的可读性,如果该文件为不可读文件,则提示无权限复制该文件。② 如果该文件为可读文件,出现提示框提示复制操作时确定还是取消。如果取消,则返回当前位置。如果点击确定,就复制文件。当用户点击文件或者文件夹时触发 onItemLongClick 事件。当判断为长按操作就是复制操作,实现代码如下。

```
public boolean onItemLongClick(AdapterView<?> parent, View view, int position, long id){
  if (isAddBackUp == true) {
      if (position! = 0 && position! = 1){ //如果返回上一级的两列存在才能进行复制
           itemLongClickListener(new File(lFilePaths.get(position)));}}
    if (lCurrentFilePath.equals(lSDCard)){ //判定复制文件时的操作环境
      itemLongClickListener(new File(lFilePaths.get(position)));} //转入长按操作事件
    return false;}
```

长按操作触发的 itemLongClickListener 事件,代码如下。

```
private void itemLongClickListener(final File file) {
    if (file.canRead()) {copyDialog(file); //文件可读才可进行复制,并调用 openDialog 模块
    } else { //文件不可读,不允许复制
        Toast.makeText(this, "无权限复制该文件!", Toast.LENGTH_SHORT).show();}}
```

在 MainActivity 中,新建一个模块 opyDialog,用于实现文件的复制操作。

```
private void copyDialog(final File file) {
  new AlertDialog.Builder(MainActivity.this)
  .setTitle("提示!").setMessage("是否要复制该文件?")
  .setPositiveButton("确定", new DialogInterface.OnClickListener() {
  public void onClick(DialogInterface dialog, int which) {
     isCopy = true; //设置复制标志位
     mCopyFileName = file.getName(); //设置复制文件名
     mOldFilePath = lCurrentFilePath + File.separator + mCopyFileName; //设置文件路径
     Toast.makeText(MainActivity.this,"已复制!",Toast.LENGTH_SHORT).show();
     initFileListInfo(file.getParent()); //复制之后重新遍历该文件的父目录} })
  .setNegativeButton("取消", null).show();}
```

文件复制的运行效果如图 12-6 所示。

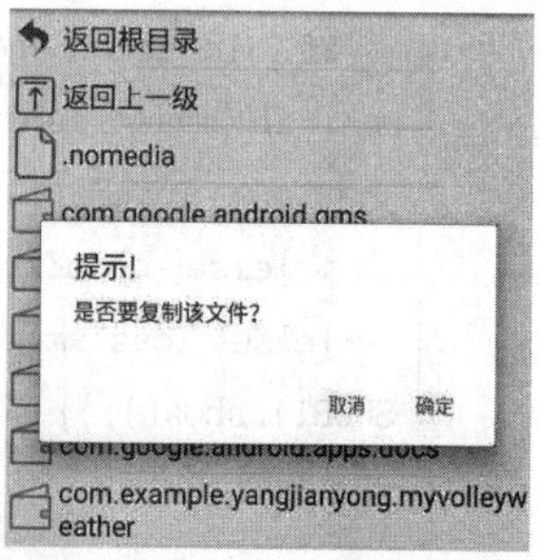

图 12-6 复制操作效果图

12.5.4 粘贴功能实现

粘贴功能的实现步骤如下:(1) 当用户点击粘贴按钮,判断操作路径。(2) 如果在同一文件夹下,则粘贴操作无效,不执行。(3) 如果不在原文件夹下,则复制文件,粘贴成功,目录重新更新。实现代码如下。

```
private void onPaste() { //判断文件复制和粘贴路径是否不同,路径不同才能进行粘贴操作
    if (!mOldFilePath.equals(mNewFilePath) && isCopy == true) {
        mNewFilePath = lCurrentFilePath + File.separator + mCopyFileName;
        copyFile(mOldFilePath, mNewFilePath); //调用复制方法
        initFileListInfo(lCurrentFilePath); //刷新当前目录文件列表
    }else{Toast.makeText(MainActivity.this, "未复制文件!",Toast.LENGTH_LONG).show(); }}
```

其核心的操作是 copyFile，代码如下。

```
public int copyFile(String fromFile, String toFile) {
    File[] currentFiles;File from = new File(fromFile); //获得要复制文件的路径
    if (from.isDirectory()) { //当复制的是文件夹的情况下,则获取当前目录下的全部文件
        currentFiles = from.listFiles();File targetDir = new File(toFile);
        if (!targetDir.exists()){targetDir.mkdirs(); //如果目标位置没有目录,则创建目录}
        for (int i = 0; i < currentFiles.length; i++) { //遍历要复制目录下的全部文件
            if (currentFiles[i].isDirectory()){ copyFile(currentFiles[i].getPath() +"/",
toFile + "/"+ currentFiles[i].getName() + "/"); //是文件就直接进行复制粘贴操作
            } else { CopySdcardFile (currentFiles [i]. getPath ( ), toFile + "/" +
currentFiles[i].getName()); //有子目录就进行递归} }
    } else {CopySdcardFile(fromFile, toFile); //是 SD 的情况下交由 CopySdcardFile 处理}
return 0;}
```

SD 卡文件的复制操作由 CopySdcardFile 函数完成，代码如下。

```
public int CopySdcardFile(String fromFile, String toFile) {
    try {
        InputStream fos = new FileInputStream(fromFile); //需要复制文件目录的输入流
      OutputStream to = new FileOutputStream(toFile); //文件粘贴目录的输出流
      byte bt[] = new byte[1024]; int c;
      while((c = fos.read(bt))> 0){to.write(bt,0,c); //根据复制文件目录写入新的粘贴目录}
        fos.close(); to.close();return 0; //操作完成之后,关闭文件的操作流
    } catch (Exception ex) { return -1; } }
```

12.5.5 动态权限申请功能实现

在程序中要获取手机内部的文件，需要访问手机的内存系统。这个权限属于系统权限，需要在程序运行时候获得授权，授权功能的实现分为几个步骤。

(1) 在 AndroidManifest.xml 文件中，声明权限。

```
<uses-permission android:name="android.permission.WRITE_EXTERNAL_STORAGE" />
```

(2) 需要两个模块用于获得权限功能，分别为 getPermissions()模块用于添加 SD 卡读写动态权限，onRequestPermissionsResult ( ) 模块用于动态权限的回调方法。在

MainActivity 中添加并编写这两个模块，相关代码请参考第 7 章内容。

12.5.6 MainActivity 代码实现

在 MainActivity 中先定义程序中用到的一些变量，并在 onCreate 方法中，完成权限的申请、控件的初始化、获取文件目录和设置事件监听等工作。

```
public class MainActivity extends ListActivity implements View. OnClickListener, AdapterView.
OnItemLongClickListener {
  private List < String > lFileName = null; //保存显示文件列表的名称
  private List < String > lFilePaths = null; //保存显示的文件列表的相对应的路径
  private String lSDCard = Environment. getExternalStorageDirectory(). toString(); //SD 卡
  LinearLayout exit_layout, paste_layout;
  private final int SDK_PERMISSION = 1; //申请权限
  private int ACTIVITY_VIEW_ATTACHMENT = 0;
  @Override
 protected void onCreate(Bundle savedInstanceState) {
        super. onCreate(savedInstanceState); setContentView(R. layout. activity_main);
        if (ContextCompat. checkSelfPermission(this, "android. permission. WRITE_EXTERNAL_
STORAGE")! = PackageManager. PERMISSION_GRANTED) { getPersimmions();
        } else { initView(); //调用获取相关控件方法
              initFileListInfo(lSDCard); //显示 SD 卡目录文件
              getListView(). setOnItemLongClickListener(this);} }
              ………… //各个模块 }
```

12.5.7 项目配置

在 app/res 中创建 xml 文件夹，新建一个 xml 文件，命名为 file_paths. xml。

```
< paths >< external-path path = "Android /data /com. yangjianyong. myfilecopyandpaste /" name = "
files_root" />< external-path path = "." name = "external_storage_root" />< /paths >
```

在 AndroidManifest. xml 中添加 URI 权限，代码如下。

```
< provider android:name = "android. support. v4. content. FileProvider"
    android:authorities = "com. yangjianyong. myfilecopyandpaste"
    android:exported = "false"  android:grantUriPermissions = "true">
    < meta-data android:name = "android. support. FILE_PROVIDER_PATHS"
          android:resource = "@xml /file_paths" /> < /provider >
```

这样，整个项目就可以流畅地运行了。

【微信扫码】
第 12 章
相关文件

小结

通过本章的学习,一个 Android 本地文件管理的 App 项目完成了。通过这个项目,我们将前面学过的知识进行了一定的整合提高,包括嵌套布局文件设计、动态权限申请、Intent 的调用、数据存储等,这是对开发能力的一种提升。尽管安卓市场中,类似文件管理有很多同类型的软件,但是通过自己的开发,可以实现自己的设想,按照自己的要求定制功能,并最终能把项目运行起来,这样的感觉的确很不错。这也是 IT 开发者的一种快乐。

第 13 章

Android 项目开发:星座运势 App

前一章实现了一个基于 Android 的文件管理 App。在这个 App 中所用到的数据,都是由开发者或用户提供,事先存储在设备的内存或 SD 卡中,属于静态数据。但在实际的开发中,我们需要获取和处理实时的、动态的数据,比如快递物流的数据、车牌的数据、天气的数据等。那么我们可以从哪里获得这些数据? 这些数据在 Android 程序的开发中如何调用? 本章我们将结合网络提供的数据接口服务,使用动态数据来实现一个星座运势的 App。

13.1　需求分析和可行性分析

在正式开发之前,首先要确定项目应该具备的功能,并将其罗列和整理出来,开发者才好对照要求去一一实现。经过需求分析,我们认为一个星座运势的手机 App 需要具备的功能如下:① 用户可以输入自己的出生日期,获得对应的星座。② 可以查看本人星座对应的今天、明天、本周的运势。③ 可以自由地切换查看星座对应的本月本年的运势。虽然看上去只有几个主要的功能,但是要全部实现这些功能,需要用到网络、数据格式转换、UI、数据存储和服务等技术,非常考验开发者的综合应用能力。

向大家推荐一个不错的网上数据 API——极速数据,官方地址为 https://www.jisuapi.com/,如图 13－1 所示,这个数据接口能够满足系统的开发要求。使用之前必须先点击页面右上角的“注册”按钮进行注册。注册的过程比较简单,这里就不一一介绍了。成功注册后会分配一个 AppKey,这个比较重要,请保存好。通过登录可以进入自己的 App 数据管理界面,如图 13－2 所示。

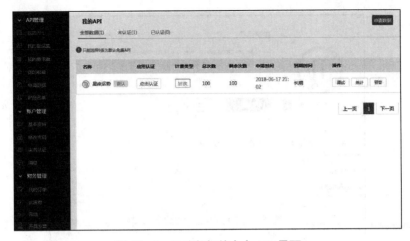

图 13－1　极速数据的主界面

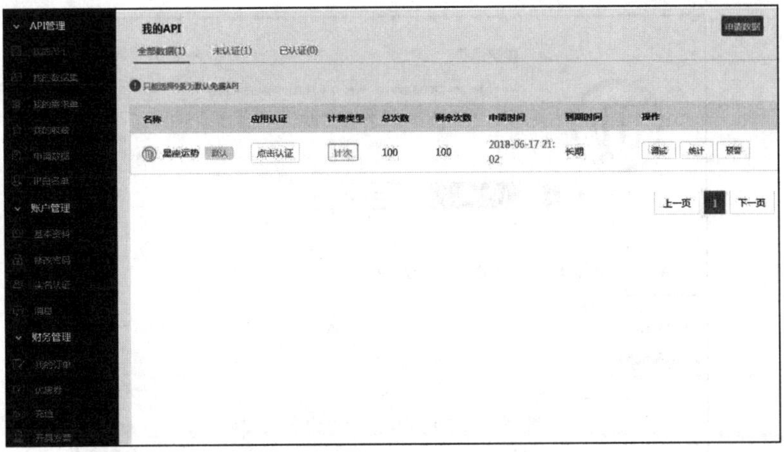

图 13－2　极速数据的个人 API 界面

在“我的 API”选项卡中，点击右侧的“申请数据”按钮，进入申请 API 界面，如图 13－3 所示。首先选择“星座算命”，然后在下面选择“星座运势”，很快就能够通过数据申请。这个数据接口申请是免费的，每天允许 100 次调试，对开发者而言，基本足够了。

图 13－3　申请数据界面

13.2　系统分析和设计

网络数据和接口申请成功之后，不代表着马上可以开始系统的开发。“磨刀不误砍柴工”，项目开发之前，对系统进行分析和良好的设计和规划是至关重要的。

运行的系统中，如何显示数据是比较关键的，因此我们首先要对网络提供的接口和数据进行分析。点击申请好的 API 名称查看数据信息，如图 13－4 所示。

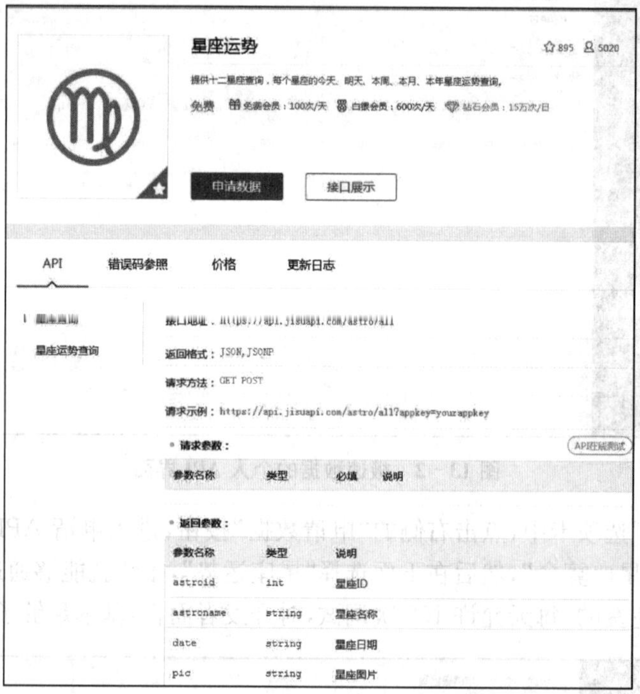

图 13-4　星座查询 API 接口说明

接口地址 https://api.jisuapi.com/astro/all 表示星座数据的访问接口。在浏览器中访问，需要使用请求示例中地址 https://api.jisuapi.com/astro/all? Appkey=yourAppkey 访问，其中 yourAppkey 是在申请系统账号成功之后，系统分配的 AppKey。

在浏览器中输入地址 https://api.jisuapi.com/astro/all? Appkey=myAppkey，获得如图 13-5 所示网页，看起来非常乱，不像通常我们看到的网页。注意在接口说明中，明确指出返回的格式是 JSON/JSONP 格式，在前面的章节中介绍过这类数据格式，这里教给大家一个小窍门，可以将这个页面保存为 JSON 格式，然后在浏览器中打开，如图 13-6 所示。

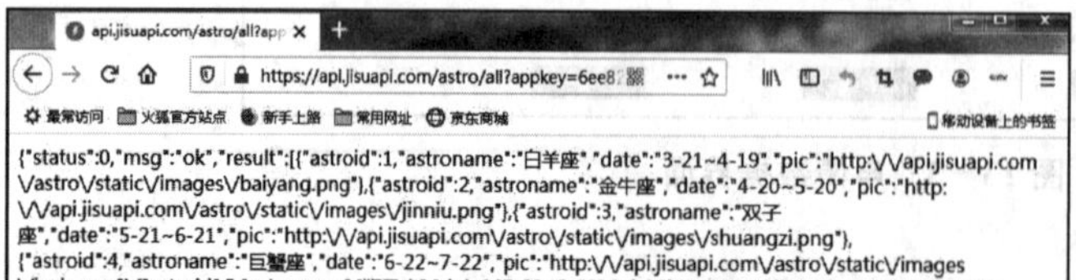

图 13-5　网页上的数据反馈

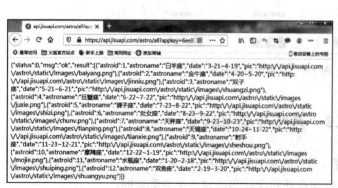

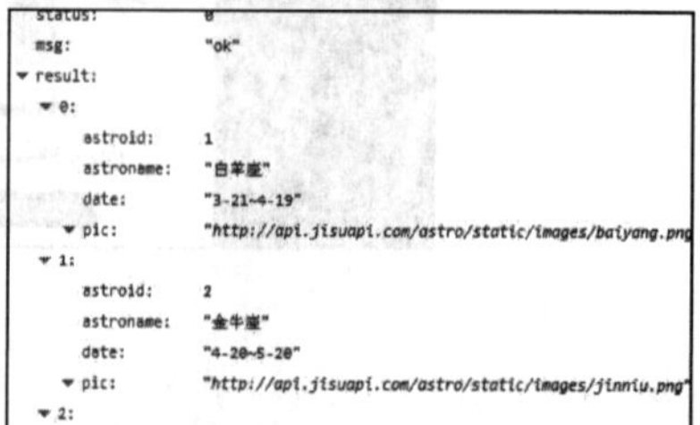

图 13-6　星座查询的 JSON 格式

整理下显示样式，我们可以看见以下格式的数据。

{"status":0,"msg":"ok","result":[{"astroid":1,"astroname":"白羊座","date":"3-21～

4-19","pic":"http:\/\/api.jisuapi.com\/astro\/static\/images\/baiyang.png"},{"astroid":2,"astroname":"金牛座","date":"4-20～5-20","pic":"http:\/\/api.jisuapi.com\/astro\/static\/images\/jinniu.png"},……]}。可以看到，这是 JSON 格式的数据信息，status 和 msg 表示数据返回正确。接下来为 JSON 格式的数组，数组的每一个元素都代表一个星座的信息，其中 astroid 为星座编号，astroname 为星座名称，date 为星座的日期范围，pic 为星座的代表图标。

在星座查询的下面是星座运势查询的 API 接口说明，如图 13-7 所示。

图 13-7　星座运势查询 API 接口说明

这个接口说明和上面的差不多，需要注意的是请求示例 https://api.jisuapi.com/astro/fortune? astroid=1&date=2016-01-19&Appkey=yourAppkey 中几个参数的用法。astroid=1 表示首先要获得星座的 astroid 值，然后代入当天的日期 date=2016-01-19，最后配上开发者的 AppKey，这样就可以发送请求获得星座运势的信息。尝试访问网址 https://api.jisuapi.com/astro/fortune? astroid=9&date=2020-01-20&Appkey=6********** 以获得星座运势信息，获得页面如图 13-8 所示。采用上面同样的方法，保存为 JSON 格式，使用浏览器打开，显示如图 13-9 所示，可以看见各个日期和时间段的运势信息。

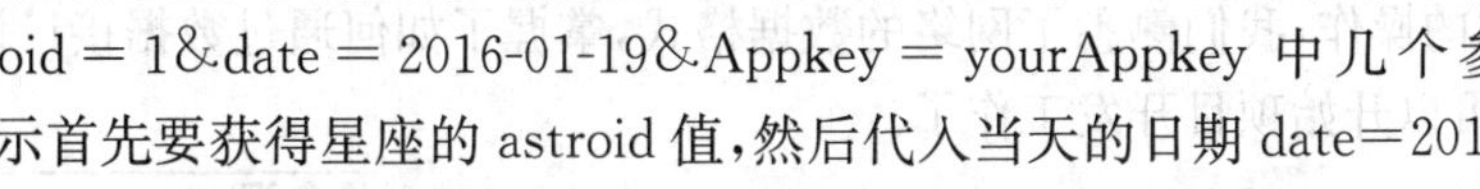

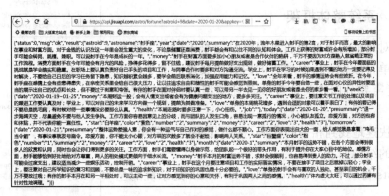

图 13-8　星座运势反馈网页

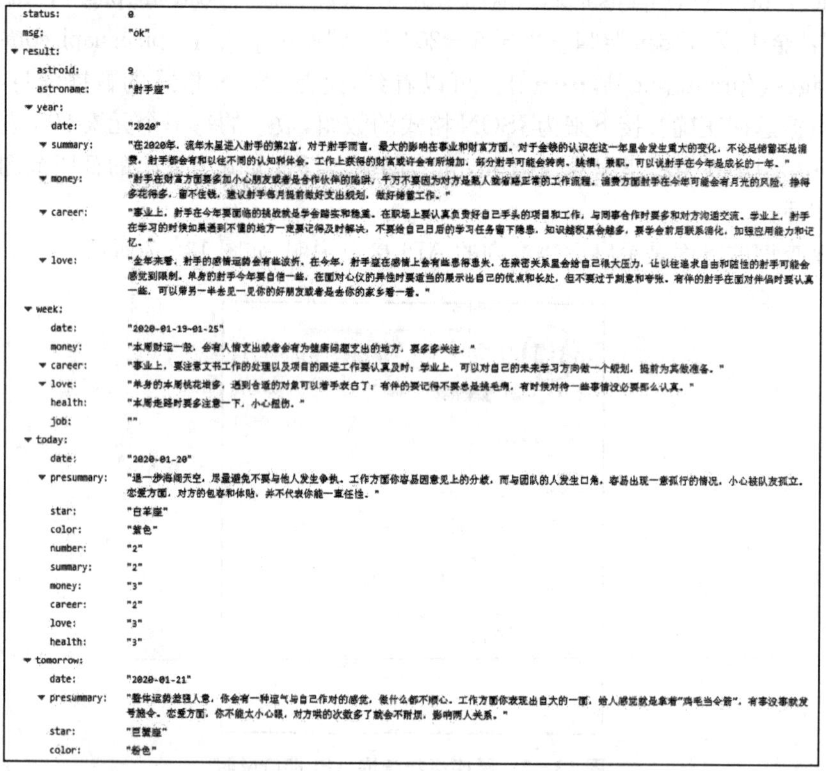

```
status:         0
msg:            "ok"
▼ result:
    astroid:        9
    astroname:      "射手座"
  ▼ year:
      date:         "2020"
    ▼ summary:      "在2020年，流年木星进入射手的第2宫，对于射手而言，最大的影响在事业和财富方面。对于金钱的认识在这一年里会发生重大的变化，不论是储蓄还是消费，射手都会有和以往不同的认知和体会。工作上获得的财富或许会有所增加，部分射手可能会跨界、跳槽、兼职。可以说射手在今年是成长的一年。"
    ▼ money:        "射手在财富方面要多加小心朋友或者是合作伙伴的陷阱，千万不要因为对方是熟人就省略正常的工作流程。消费方面射手在今年可能会有月光的风险，挣得多花得多，留不住钱，建议射手每月提前做好支出规划，做好储蓄工作。"
    ▼ career:       "事业上，射手在今年要面临的挑战就是学会踏实和稳重。在职场上要认真负责好自己手头的项目和工作，与同事合作时要多和对方沟通交流。学业上，射手在学习的时候如果遇到不懂的地方一定要记得及时解决，不要给自己日后的学习任务留下隐患，知识越积累会越多，要学会前后联系消化，加强应用能力和记忆。"
    ▼ love:         "全年来看，射手的感情运势会有些波折。在今年，射手座在感情上会有些患得患失，在亲密关系里会给自己很大压力，让以往追求自由和随性的射手可能会感觉到限制。单身的射手今年要自信一些，在面对心仪的异性时要适当的展示出自己的优点和长处，但不要过于刻意和夸张。有伴的射手在面对伴侣时要认真一些，可以带另一半去见一见你的好朋友或者是去你的家乡看一看。"
  ▼ week:
      date:         "2020-01-19~01-25"
      money:        "本周财运一般，会有人情支出或者会有为健康问题支出的地方，要多多关注。"
    ▼ career:       "事业上，要注意文书工作的处理以及项目的跟进工作要认真及时；学业上，可以对自己的未来学习方向做一个规划，提前为其做准备。"
    ▼ love:         "单身的本周桃花增多，遇到合适的对象可以着手表白了；有伴的要记得不要总是挑毛病，有时候对待一些事情没必要那么认真。"
      health:       "本周走路时要多注意一下，小心扭伤。"
      job:          ""
  ▼ today:
      date:         "2020-01-20"
    ▼ presummary:   "退一步海阔天空，尽量避免不要与他人发生争执。工作方面你容易因意见上的分歧，而与团队的人发生口角，容易出现一意孤行的情况，小心被队友孤立。恋爱方面，对方的包容和体贴，并不代表你能一直任性。"
      star:         "白羊座"
      color:        "紫色"
      number:       "2"
      summary:      "2"
      money:        "3"
      career:       "2"
      love:         "3"
      health:       "3"
  ▼ tomorrow:
      date:         "2020-01-21"
    ▼ presummary:   "整体运势差强人意，你会有一种运气与自己作对的感觉，做什么都不顺心。工作方面你表现出自大的一面，给人感觉就是拿着"鸡毛当令箭"，有事没事就发号施令。恋爱方面，你不能太小心眼，对方哄的次数多了就会不耐烦，影响两人关系。"
      star:         "巨蟹座"
      color:        "粉色"
```


图 13－9　运势查询的 JSON 格式

通过上述的操作，我们熟悉了网络的数据格式，掌握了如何通过数据接口访问并获得数据，接下来就可以开始项目开发工作了。

13.3　系统基础模块实现

在 Android Studio 中新建一个项目，命名为 Project_Teach_AstroFortune。

13.3.1　系统整体框架

一个相对完整的项目和演示程序是完全不同的。演示程序相对简单，代码不多，但项目的内容比较多。为了便于维护，开发者需要分门别类地存放各种类型的资源和代码文件。

为了更快速地找到相关的代码，按照代码类型分类，整个项目代码的结构如下。

manifests 目录下的 AndroidManifest.xml 是系统的全局配置文件。

java 包下主要是代码文件，和界面相关的代码放在

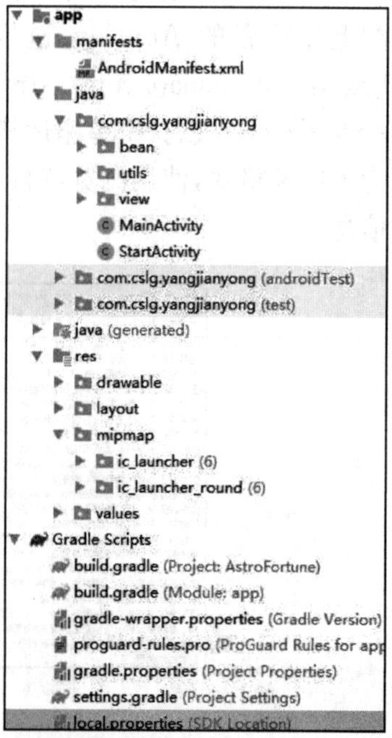

图 13－10　项目的代码结构

view 包下，网络和数据处理的代码放在 utils 包下，数据封装代码放在 bean 包下。

资源文件统一在 res 包下，图片资源文件放在 res/drawable 中，布局文件放在 res/layout 中，字符串、颜色和样式文件在 res/values 中。

之前的案例，一开始都会进入界面设计，但是在涉及网络数据的 App 中，首先应该实现网络功能部分，这样界面设计才更有目的性。

13.3.2　基础功能实现文件

从上面的介绍中，我们知道数据接口返回的数据类型，这些数据在 App 中能够使用，必须要对其进行数据封装，封装为实体类。这样的实体类除了要包含字段之外，还要有读写属性。图 13－11 所示是星座数据的返回参数信息。

• 返回参数：

参数名称	类型	说明
astroid	int	星座ID
astroname	string	星座名称
date	string	星座日期
pic	string	星座图片

图 13－11　星座数据的返回参数

根据返回的参数信息，编写参数的实体类，在 src 中新建一个文件夹 bean，用于存放项目用到的实体类代码，在里面新建一个 java 类，名称为 AstroBean。代码中包含 astroid、astroname、date、pic 字段，还有字段的 get/set 方法，构成实体类中的读写属性。同理还有星座运势的返回参数，也需要实现同样的实体类，这里就不再详述，请读者自行完成。

13.3.3　项目工具类实现

在 App 中需要连接网络，接收数据，处理 JSON 格式等，这些功能相对独立，可复用性高，为了便于管理，将这些功能代码全部放置到 utils 包中。

首先实现数据网址的查询功能，在 utils 包中新建一个接口，命名为 NetContantStr。对照星座查询和星座运势的请求网址格式 https://api.jisuapi.com/astro/all? Appkey＝yourAppkey 和 https://api.jisuapi.com/astro/fortune? astroid＝1&date＝2016－01－19&Appkey＝yourAppkey，发现前面基本一致，后面根据不同的查询目的和参数有差异，将查询的网址拆分为一些独立的字符串，这样便于组合搭配。

```
public interface NetContantStr {
    String APP_KEY = "Appkey = *****(你的 Appkey)"; String ASTROID_URL = "fortune?
astroid = ";
    String HTTP_URL = "https: //api.jisuapi.com /astro /";
    String ASTRO_URL = "all?"; String DATE_URL = "date = ";}
```

在 utils 包中新建一个类名为 AstroUtils 用于查询星座信息，代码如下。

```
public class AstroUtils {
  public int astroid;
  public int[] astroEdgeDay = { 0, 20, 19,21, 20, 21, 22, 23, 23,23, 24, 23, 22 };
  //根据用户提供的日期数据获得星座的 id
  public int astroidByDate(int month, int day) {
    if ((month = = 1) || month = = 2 || (month = = 3 && day <= 20)) {
      if (day> = astroEdgeDay[month]) {astroid = month + 10;}else astroid = month + 9;}
      else if (day >= astroEdgeDay[month]){astroid = month - 2;}
            else astroid = month - 3; return astroid;}
  //检查用户提供的日期是否正确
  public boolean monthDay(int month, int day){
    switch(month){
      case 1:case 3:case 5:case 7:case 8:case 10:case 12: {
              if (day> 0 && day< = 31) return true;  else return false;}
      case 4:case 6:case 9:case 11:{
              if (day> 0 && day< 31) return true;else return false;}
      case 2:{ if (day> 0 && day< 30)  return true;else return false; }
      default:return false;}}
  //根据传入的地址获得星座的返回数据 InputStream
  public InputStream getAstroInputStream() throws Exception {
    String url = NetContantStr.HTTP_URL + NetContantStr.ASTRO_URL + NetContantStr.APP_KEY;
    HttpURLConnection conn =  (HttpURLConnection) new URL(url).openConnection();
    conn.setReadTimeout(5000);conn.setConnectTimeout(10000);
    int responseCode = conn.getResponseCode(); return conn.getInputStream(); }
  public String loadAstroData() throws Exception {
    InputStream is = getAstroInputStream();
    String json = TextConvertUtils.inputstreamToString(is);
    if (is! = null){ is.close(); } return  json;}}
```

有了星座的信息，如何根据星座的信息查询对应的星座运势的信息呢？

在 utils 包中，新建一个名为 FortuneUtils 的类，代码如下。

```
public class FortuneUtils {
    public  InputStream getAstroFortuneInputStream(int astroid, String date) throws Exception {
        String url = NetContantStr.HTTP_URL + NetContantStr.ASTROID_URL + astroid + "&" +
NetContantStr.DATE_URL + date + "&" + NetContantStr.APP_KEY;
        HttpURLConnection conn = (HttpURLConnection)new URL(url).openConnection();
        conn.setReadTimeout(5000);conn.setConnectTimeout(10000);
        int responseCode = conn.getResponseCode(); return conn.getInputStream(); }
    public String loadFortuneData(int astroid, String date) throws Exception {
    InputStream is  =  getAstroFortuneInputStream(astroid,date);
    String json  =  TextConvertUtils.inputstreamToString(is);
    if (is! = null){is.close(); } return json; }}
```

这部分代码比较简单,就是根据传入的星座的 ID 值,获得对应的星座运势的 JSON 数据。

在星座和运势的查询中,采用 InputStream 方式,这种方式返回的数据如果要进一步的处理,需要进行数据格式的转换,转换为 String 类型,在 utils 包中新建一个类,命名为 TextConvertUtils,专门用于数据的传唤处理。代码如下。

```
public class TextConvertUtils {
public static String inputstreamToString(InputStream is) throws IOException {
   BufferedReader reader = new BufferedReader(new InputStreamReader(is, "UTF - 8"));
   StringBuilder builder = new StringBuilder();String line;
   while ((line = reader.readLine()) ! = null) {builder.Append(line).Append('\n'); }
     reader.close();return builder.toString(); } }
```

13.4　系统界面实现

在项目中,设计和构建 App 的界面是一件非常重要的工作。界面的好坏直接影响 App 品质,也关乎能否吸引更多用户。在这个 App 中,根据系统的需求,通过合理地规划和设计,将项目的界面分为启动界面和信息显示界面两大块。

13.4.1　启动界面

项目启动之后,首先要获得用户的生日信息。

在 App 的 res/layout 下,新建一个 layout 文件,命名为 Astro.xml,修改代码如下。

```
<androidx.constraintlayout.widget.ConstraintLayout
      xmlns:android = "http://schemas.android.com/apk/res/android"
      android:layout_width = "match_parent" android:layout_height = "match_parent">
      <RelativeLayout android:background = "@drawable/background"
            android:layout_width = "match_parent"
            android:layout_height = "match_parent">
            <TextView  android:layout_width = "match_parent"
                  android:layout_height = "wrap_content"
                  android:gravity = "center_horizontal"
                  android:layout_marginTop = "20dp"  android:text = "请填写日期"
                  android:textSize = "40dp"   android:textColor = "#FFFFFF" />
            <EditText android:id = "@+id/edit_month"
               android:layout_width = "60dp" android:layout_height = "wrap_content"
               android:layout_marginTop = "200dp" android:layout_marginLeft = "80dp"
               android:textSize = "40dp" android:textColor = "#FFFFFF" />
            <TextView  android:id = "@+id/txtmonth"
                  android:layout_width = "50dp" android:layout_height = "wrap_content"
                  android:layout_alignBaseline = "@id/edit_month"
                  android:layout_toRightOf = "@id/edit_month" android:text = "月"
```

```
            android:textColor = "#FFFFFF" android:textSize = "40dp" />
        <EditText  android:id = "@+id/edit_day" android:layout_width = "60dp"
            android:layout_height = "wrap_content" android:textSize = "40dp"
            android:layout_marginTop = "200dp" android:textColor = "#FFFFFF"
            android:layout_toRightOf = "@id/txtmonth" />
        <TextView  android:id = "@+id/txtday"  android:layout_width = "50dp"
            android:layout_height = "wrap_content" android:textColor = "#FFFFFF"
          android:layout_alignBaseline = "@id/edit_day" android:textSize = "40dp"
            android:layout_toRightOf = "@id/edit_day" android:text = "日" />
        <Button android:id = "@+id/getAstroButton"
            android:layout_marginTop = "450dp"
            android:layout_width = "wrap_content"
            android:layout_height = "wrap_content"
            android:textColor = "#FFFFFF" android:layout_marginLeft = "80dp"
            android:background = "@drawable/border_line_while"
            android:textSize = "30dp" android:text = "星座运势" />
    </RelativeLayout>
</androidx.constraintlayout.widget.ConstraintLayout>
```

启动界面比较简单，提供两个输入框，让用户输入自己的出生日期。提供一个 Button，让用户点击之后进行跳转。

13.4.2 信息显示主界面实现

当用户在启动界面输入个人出生日期，点击按钮跳转之后，正常情况下，程序将获得星座运势的返回信息。返回的信息包括今日、明日、本周等多种数据信息。为了让用户能够一目了然地查看个人的星座运势信息，信息显示界面需要具备展现清晰、浏览方便等特点。

手机的屏幕都比较小，项目中需要浏览的信息又比较多，一个屏幕容纳不下，因此需要一个可以上下滚动的显示视图。这里使用 VerticalScrollView 组件，这个组件可以实现屏幕内容的上下滚动。在 app/src 中新建一个包，命名为 view，在包中新建一个类，命名为 VerticalScrollView，修改代码如下。

```
public class VerticalScrollView extends ScrollView {
  public VerticalScrollView(Context context) { super(context); }
  public VerticalScrollView(Context context, AttributeSet attrs) {super(context, attrs);}
  public VerticalScrollView(Context context, AttributeSet attrs, int defStyleAttr){
        super(context, attrs, defStyleAttr);}
  public VerticalScrollView (Context context, AttributeSet attrs, int defStyleAttr, int
defStyleRes) {
     super(context, attrs, defStyleAttr, defStyleRes);}
     private float mDownPosX = 0;private float mDownPosY = 0;
     @Override
```

```
public boolean onInterceptTouchEvent(MotionEvent ev) {
  final float x = ev.getX();final float y = ev.getY();final int action = ev.getAction();
  switch (action) {
    case MotionEvent.ACTION_DOWN:mDownPosX  =  x;mDownPosY  =  y;break;
    case MotionEvent.ACTION_MOVE:final float deltaX = Math.abs(x - mDownPosX);
          final float deltaY  =  Math.abs(y  -  mDownPosY);
          if (deltaX > deltaY) {   return false; }}
    return super.onInterceptTouchEvent(ev);}}
```

VerticalScrollView 继承于 ScrollView,可以实现屏幕内容的上下滚动,类中方法的代码直接使用即可,需要修改的是 onInterceptTouchEvent 中的代码。

在 VerticalScrollView 的基础上,可以实现基于 VerticalScrollView 的布局文件。在 res/layout 下,新建一个 layout 文件,命名为 activity_main.xml,修改代码如下。

```
<com.cslg.yangjianyong.view.VerticalScrollView android:id = "@ + id /activity_main" …… >
  <RelativeLayout android:orientation = "vertical" …… >
    <TableLayout android:id = "@ + id /tablelayout" ……. >
      <TableRow android:layout_height = "wrap_content" …… >
        <TextView android:id = "@ + id /txtastroname" ……  />
        <TextView android:id = "@ + id /astroname" ……  />
      </TableRow>
      <TableRow android:layout_height = "60dp" …… >
        <TextView android:id = "@ + id/ txtdate" ……  />
        <TextView android:id = "@ + id /astrodate" ……  />
      </TableRow>
    </TableLayout>
    <ImageView android:id = "@ + id /imageastro" ……  />
    <Button android:id = "@ + id /btntodayfortune" android:layout_height = "40dp"
      android:layout_marginTop = "80dp" android:layout_width = "63dp"
      android:textColor = "#FFFFFF" android:text = "今日运势"
      android:background = "@drawable /border_line_while" android:textSize = "13sp"  />
      …… //以此类推完成明日运势、本周运势、本月运势、本年运势按钮
     <LinearLayout android:layout_width = "match_parent"
         android:layout_height = "wrap_content" android:padding = "5dp"
         android:layout_marginTop = "140dp"
         android:orientation = "vertical"android:background = "#56000000">
     </LinearLayout>
     <ProgressBar android:layout_width = "wrap_content"
         android:layout_height = "wrap_content"   android:id = "@ + id /progress"
         android:layout_centerInParent = "true"   android:visibility = "gone"  />
  </RelativeLayout>
</com.cslg.yangjianyong.view.VerticalScrollView>
```

整个布局文件实现的界面可以实现上下翻屏。效果如图 13 - 12 所示。

图 13 - 12 App 信息显示主界面

13.4.3 详细信息显示界面设计

在图 13 - 12 中点击每个运势的按钮，需要显示对应的运势。各个运势显示的内容不完全一致，所以需要根据返回的数据分别设计。

在 res/layout 下新建若干个 layout 文件，命名为 today.xml、tomorrow.xml、week.xml、month.xml、year.xml 分别对应今日、明日、本周、本月和本年。

today.xml 文件中，使用 TextView 和 RatingBar 等显示文字和星级评分等信息代码如下。

```
< androidx.constraintlayout.widget.ConstraintLayout …… >
< LinearLayout android:orientation = "vertical" …… >
        < TextView android:id = "@ + id /todaydate" style = "@style /text_fortune" android:
text = "日期" />< include layout = "@layout /line" />
    …… //贵人星座、幸运颜色、幸运数字等参照日期 TextView 设计
    < TextView android:id = "@ + id /todaymoney" android:text = "财运运势"  …… />
    < RatingBar android:id = "@ + id /rbtodaymoney" android:numStars = "5" …… />
    < include layout = "@layout /line" />
    …… //综合、工作、爱情、健康运势等参照财运运势 TextView + RatingBar 设计
< /LinearLayout >
< /androidx.constraintlayout.widget.ConstraintLayout >
```

其他如明日、本周等运势的界面请参照今日运势的界面，这里不再一一详细介绍。

编写好的各个详解信息界面，通过 include 嵌套方式，将布局文件嵌套到主布局中，例如 today.xml 嵌入 activity_main.xml 方式，代码如下。

```
< LinearLayout android:id = "@ + id /todayfortune" ……>
    < TextView android:id = "@ + id /today"  android:text = "今日运势" …… />
    < include layout = "@layout /today" />
< /LinearLayout >
```

实现的效果如图 13 - 13 所示，其他如明日、本周等运势的详解界面也请读者参照编写。

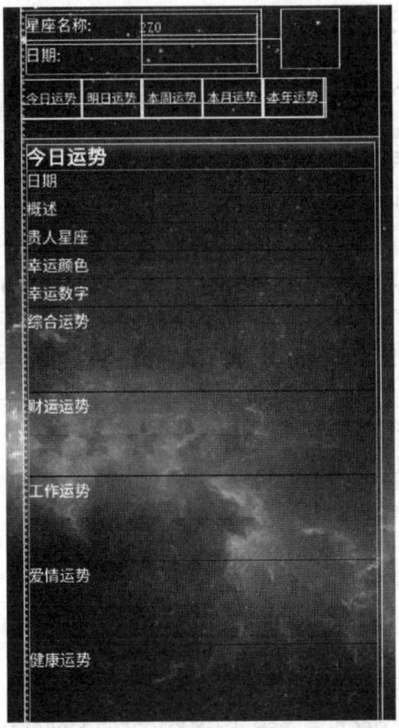

图 13 - 13　App 信息显示主界面

13.5　系统核心功能实现

在 app/src 中,新建一个 java 类,命名为 MainActivity.java。它的作用就是实现界面的数据显示,是整个 App 的核心功能。在代码中将需要实现的功能分为几个模块。

13.5.1　图片下载和控件初始化功能

showImage()用于实现每个星座图片的下载和显示,代码如下。

```
private void showImage(){
    final Handler handler = new Handler();
    new Thread() {
      public void run() {
          bm = new ApacheHttpClient().getHttpBmp(url);
```

```
          handler.post(new Runnable(){
              @Override
              public void run() { imageastro.setImageBitmap(bm);}
          });}}.start();}
```

initView()方法用于初始化项目中用到的全部控件,这里使用"……"代替其他控件初始化。

```
private void initViews() { astroname = findViewById(R.id.astroname);……}
```

13.5.2 线程和控件数据更新

hideProgressBarOnUiThread()用于实现进度条的更新，这里使用了线程模式。

```
private void hideProgressBarOnUiThread() {
    runOnUiThread(new Runnable() {
        @Override
        public void run() { progressBar.setVisibility(View.GONE); }  });}
```

loadAstroFortune()方法用于根据星座 ID 和日期获得运势数据，并绑定到控件上。

```
private void loadAstroFortune(int astroId, String Date) {
    progressBar.setVisibility(View.VISIBLE);
    new Thread(new Runnable() {
      @Override
   public void run() {
   try {
     String jsonAstro = new AstroUtils().loadAstroData();
     JSONObject jsonAstroObject = new JSONObject(jsonAstro);
     final JSONArray jsonAstroResult = jsonAstroObject.optJSONArray("result");
     final JSONObject jsonastro = (JSONObject)jsonAstroResult.opt(astroid - 1);
     String jsonFortune = new FortuneUtils().loadFortuneData(astroid, date);
     JSONObject jsonFortuneObject = new JSONObject(jsonFortune);
     final JSONObject jsonFortuneResult = (JSONObject)jsonFortuneObject.get("result");
     ……//获得运势其他数据
     runOnUiThread(new Runnable() {
     @Override
     public void run() {
       try {
           astroname.setText(jsonastro.getString("astroname"));
           ……//将数据信息填充到控件中
         } catch (JSONException e) {e.printStackTrace();} }  });
         } catch (Exception e) {e.printStackTrace();}
         hideProgressBarOnUiThread(); } }).start();}
```

13.5.3 OnCreate 方法实现

MainActivity 中所包含的其他模块功能实现代码如下。

```
public class MainActivity extends AppCompatActivity {
    //声明控件和变量信息
    @Override
    protected void onCreate(Bundle savedInstanceState) {
    ……
```

```
initViews();
Intent intent = getIntent();astroid =  intent.getIntExtra("astroid", 1);
loadAstroFortune(astroid, date);
btntodayfortune.setOnClickListener(new View.OnClickListener(){
@Override
public void onClick(View v) { //显示今天运势的数据同时,隐藏其他运势的界面
todayfortune.setVisibility(View.VISIBLE);tomorrowfortune.setVisibility(View.GONE);
…… //省略 weekfortune、monthfortune、yearfortune 的隐藏代码);
btntomorrowfortune.setOnClickListener(new View.OnClickListener(){});
…… //省略其他三个 tnweekfortune、btnmonthfortune、btnyearfortune 单击事件);}
private void showImage(){…… }
private void initViews() {……}
private void hideProgressBarOnUiThread() {……}
private void loadAstroFortune(int astroId, String Date) {……}}
```

运行程序，结果如图 13－14 和图 13－15 所示，可以看到显示界面，功能都实现了。

图 13－14　启动界面用户输入日期

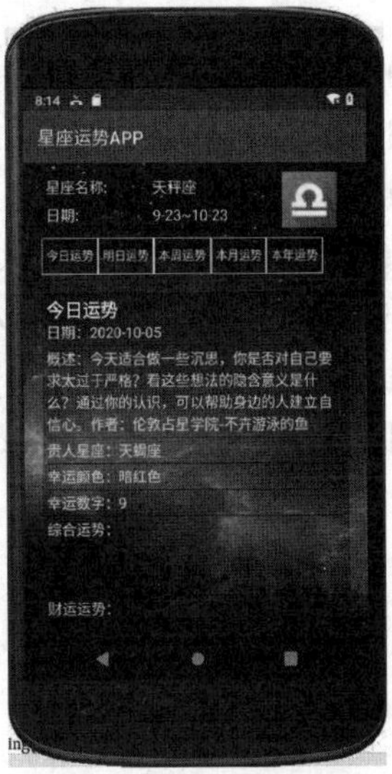

图 13－15　今日运势界面

小结

通过本章的学习，一个不算复杂的使用网络数据接口的 App 项目完成了。和前一章节的项目相比，我们对前面学习过的知识又进行了一次整合，包括界面构建、网络请求、数据格式解析、高级控件应用等。如果读者从头到尾学习完本章内容，可以看看自己具体掌握了多少知识。

【微信扫码】
第 13 章
相关文件

参考文献

[1] 李天祥.Android物联网开发细致入门与最佳实践[M].北京：中国铁道出版社，2016.

[2] 张荣，朱辉，曹小鹏.Android开发与应用[M].北京：人民邮电出版社，2010.

[3] 于连林.爱上Android[M].北京：人民邮电出版社，2017.

[4] 郭霖.第一行代码：Android[M].第二版.北京：人民邮电出版社，2016.

[5] 郭金尚.Android经典项目案例开发实战宝典[M].北京：清华大学出版社，2013.

[6] 张亚运.Android开发入门百战经典[M].北京：清华大学出版社，2017.

[7] 张伦.Android物联网应用开发[M].广州：中国财富出版社，2017.

[8] 明日学院.Android开发从入门到精通：项目案例版[M].北京：中国水利水电出版社，2017.

[9] 郑丹青.Android模块化开发项目式教程[M].北京：人民邮电出版社，2018.

[10] 肖正兴.Android移动应用开发[M].北京：中国铁道出版社，2018.

[11] 张思民.Android应用程序设计[M].第二版.北京：清华大学出版社，2018.

[12] 何富贵.新编Android应用开发从入门到精通[M].北京：机械工业出版社，2018.

[13] 许超.Android项目开发实战教程[M].北京：化学工业出版社，2018.

[14] 巫湘林.Android应用开发基础教程[M].北京：中国水利水电出版社，2017.

[15] 郑阿奇.Android实用教程：基于Android Studio[M].北京：电子工业出版社，2017.